A GEAR
CHRONOLOGY

A GEAR CHRONOLOGY

Significant Events and Dates Affecting Gear Development

William P. Crosher

Copyright © 2014 by William P. Crosher.

Library of Congress Control Number:		2014916659
ISBN:	Hardcover	978-1-4990-7115-3
	Softcover	978-1-4990-7114-6
	eBook	978-1-4990-7119-1

All rights reserved. No part of this book may be reproduced or transmitted in any form or by any means, electronic or mechanical, including photocopying, recording, or by any information storage and retrieval system, without permission in writing from the copyright owner.

Any people depicted in stock imagery provided by Thinkstock are models, and such images are being used for illustrative purposes only. Certain stock imagery © Thinkstock.

This book was printed in the United States of America.

Rev. date: 09/16/2014

To order additional copies of this book, contact:
Xlibris LLC
1-888-795-4274
www.Xlibris.com
Orders@Xlibris.com
549676

Contents

Chapter 1	Mathematics -- Metals – Paper-- Machines – Calendar -- Gears	7
Chapter 2	Standards - Metals - P/M - Mathematics – Machines - Paper – Gears - Mills	17
Chapter 3	Gear Technology from the 14th to the 17th CENTURY	35
Chapter 4	Progress Through the 18th Century	77
Chapter 5	Gear Development and Growth In The Nineteenth Century	103
Index of Individuals		243
Index of Companies		255

Chapter 1

Mathematics -- Metals – Paper-- Machines – Calendar -- Gears

The B.C. Era: Over the 6-5000 years B.C. there would be major improvements in agricultural methods, the understanding of mathematics, advancements in the use of metals and in maintaining written records. From the earliest known wheel, discovered in Kish on the banks of the Euphrates and believed to be 5,000 years old wheels with teeth have evolved. The invention of the wheel has been credited to the Elamites, the country of Elam being adjacent to the river Tigris Elamite sculptures illustrating the wheel are the basis for the claim. The Elamite wheel was from a chariot and consisted of three pieces clamped together with copper, and included a hub and tires of similar material.

These developments were followed by a period when the bronze-age became the iron-age. During the dated period 3300-2000 B.C. the Sumerian civilization in Southern Mesopotamia are credited with inventing writing, considered to be the greatest invention.

The *Vedas*, four collections of sacred Hindu literature, was written circa 1700 B.C. The poetry clearly indicates that wheeled vehicles were not in general usage.

Towards the end of the age basic machines had been established. Even a steam engine had been built by Hero in 150 B.C. The gears required for these machines included crude right angle and parallel shaft drives, but also technically advanced gearing such as the differential and gears with circular curved teeth.

Mathematics and Measurement: In Mesopotamia (present day Iraq), the birthplace of civilization, the systematic measurement and comparison of angles would begin in 6-5000 B.C. They used the ratios 60:1 and 24:1 that would later divide the day, hour, minute, and second. Mesopotamians used the base 60 as we would now use the base 10, dividing the circle into 360 equal parts. Though little used they knew of the decimal point. Writing was in use by the Uruk culture in the fourth millennium and served their bureaucratic needs. The writing materials were damp clay and a pointed instrument.

Further improvements in measurement occurred in 3000 B.C. A subdivisional rod was made to be used as a standard. A similar rod (circa 1300 B.C.) can be seen in the Egyptian Museum, Turin, Italy. In 2700 B.C., using a standardized measurement system the Egyptians built the Khufu Pyramid at Giza. Each side was 758 feet and total accuracy was within eight inches. Even more remarkable the four sides are square within three and a half seconds of arc, and within five and a half seconds of arc to the points of the compass. Further advances in mathematics occurred in 2000 B.C. The Babylonians had arrived at a value of $3\frac{1}{8}$ for pi, whilst the Egyptians had arrived at the value of $pi = 4(8/9)^2$. Some three hundred years later simple algebraic problems were being written on papyrus by the Egyptian Ahmose. It is believed that his algebra was founded on an earlier work dating back to 3400 B.C. The foundation for mathematics. This was followed by a period when the bronze-age became the iron-age. The date 1200 B.C. is normally considered as the boundary between the late Bronze Age (c.1550-1200 B.C.) and the Iron

Age (c.1200-c.580 B.C.).

In 1000 B.C. the Mesopotamians would introduce the sexagesimal system, sub-dividing the right angle into 90 degrees, each degree into sixty minutes and, each minute into sixty seconds. In the first millennium an alphabetical writing method came into being.

In China 770 B.C -- 446B.C. was the time of the Eastern Zhou dynasty. Plows were pulled by oxen, and iron tools were in use. Mathematics was taught in schools with the assistance of multiplication tables. In the Western Zhou dynasty the mathematician Shang Gao deduced …in a right angle triangle, when the base is three and altitude four, the hypotenuse is five.

During the Western Han Dynasty (206 B.C.--A.D. 24) the book "Nine Sections on the Mathematical Art" was written, detailing algebra, geometry, square and cubic roots, pi was to the value of three. Important earlier Chinese mathematical works are reputed to have been destroyed by the emperor Shih Huang-ti in 221 B.C. Unfortunately, important Chinese work was written on perishable materials, such as bark, bamboo, or silk. The theorem of Pythagoras, and decimal numeration were known.

Thales, who was born 640 B.C. and died in 550 B.C., founded the earliest Greek school of mathematics in 580 B.C. He established many of the basics required in teaching the geometry of triangles and straight lines. Between 430-490 B.C. Zeno of Elea, a famous Greek mathematician devised his famous motion paradoxes. "Zeno's Arguments on Motion" was edited by the mathematics historian Professor Florian Cajori, and published in the "American Mathematical Monthly" in 1915. This invaluable work on motion has been debated throughout the centuries.

Pythagoras started an analytical approach to numbers and trigonometry laying the foundation for more advanced mathematics. Hippocrates (470-410 B.C.) was the author of the first *"Elements of Geometry"*. He was the first to show that the ratio of the areas of two circles was equal to the square of their radii. He supplied geometric solutions to solving quadratic equations.

The Greek mathematician, Euclid (300 B.C.) wrote *"Elements of Geometry"* consisting of thirteen volumes. These were probably the most celebrated books written on mathematics and provided a foundation for all subsequent mathematics. The volumes were in use up to the earliest part of the twentieth century. It is believed that this was the first mathematics book. The Pythagorean Theorem was proven in Proposition 47, Book 1. He gathered together the combined works of Pythagoras, Hippocrates, Theaetetus, Eudoxus and other geometricians.

Archimedes, another famous Greek mathematician, improved on this work, using applied mathematics between 287 -212 B.C. He provided formulae for spheres, parabolas, cylinders, and accurately calculated pi. 1800 years later this work would be the basis for integrator theories. In addition Archimedes is believed to have developed the *Archimedes Screw*, and is known as the *Father of the Worm Gear*.

The Greek mathematician Eratosthenes (276-194 B.C) determined prime numbers by the so-called *Sieve of Eratosthenes*. The Greek philosopher and mathematician, Apollonius of Perga (250-220 B.C.), made a major contribution to geometry and would be known thereafter as the *Great Geometer*. He was the author of a definitive work on conic sections which would be the foundation of future teachings, and even today can only be understood by our most advanced mathematicians. (Ref: *Apollonius of Perga* edited by Sir T. L. Heath, Cambridge, 1896)

The Maya had an advanced agriculture and mathematical system far superior to the Egyptians. Mayans had discovered and made use of the zero digit, also the positional notation that would confuse European mathematicians for the next thousand years. They used the base twenty vigesimal system. Calculations involved a combination of ones denoted by dots and fives denoted by bars.

In 160 B.C. the Greek astronomer, Hipparchus, was born, and has been credited with the invention of trigonometry. (The first book on trigonometry was credited to al-Biruni in the 10th century A.D). In 95 B.C. the Egyptian astronomer, Ptolemy, or Claudius Ptolemaeus as he is also known, adopted Apollonius's solution to the problem of planet movement by using a complicated system of epicyclics, the Ptolemaic System. The objection methods Ptolemy used were an early adaptation of perspective.

Metals: The controlled use of fire was the first major breakthrough in our becoming a civilized society. Apart from the obvious provision of light and heat the physical properties of materials could be changed. We now have evidence that the heat treatment of stones occurred 72,000 years ago. The gloss levels on twenty-four pinnacle rock tools aged 164,000 years indicates that they too had been heated at high temperatures.

Copper was first worked in Iran about 5,500 B.C. Rock was heated until the copper ran out. The technique was developed to obtain other metals. The copper lacked hardness until bronze was developed about 3,000 B.C. by the addition of tin. In Mesopotamia 3000-2500 B.C. bronze alloyed with tin was used for tools and weapons. An early Biblical reference is in Genesis 4:22: "Tubal-Cain, who was an artificer of bronze and iron." By 1600 B.C. the Shang Dynasty (16th-11th centuries), also known as the Yin dynasty, used a highly developed bronze technology, and an advanced writing system that provided the first written evidence of Chinese history. The Chinese developed the piece mold technique and lost wax method in this period. They also (circa 220 B.C.) became accomplished in producing bronze having discovered the correct alloying elements. Chinese development was independent, their techniques differed from those used in Europe and the Middle East in that they relied on annealing, cold working and hammering. In the European Middle Bronze Age (2000 -1500BC.) copper from Cyprus was the most prized. Olive oil was used to smelt the copper which left fewer impurities. In 200 B.C. the famous Terra-cotta army of the Qin dynasty were made. The excavated chariots major parts were of bronze and minor parts of gold and silver. The swords were sharp enough to cut paper and were chrome plated to a thickness of ten to 15 micron. Chrome plating technology having supposedly been "invented" in Germany in 1937 and in America in 1950. The bronze swords also had 21.3% of tin providing a hardness equivalent to a tempered carbon steel. After being buried for two thousand years they were as shiny as if in new condition.

Tools were made from meteorite iron, wood, and stone. About 2,500 B.C., in the Middle East iron smelting was developed by following the ability to obtain temperatures of 1,500 degrees Celsius The Hittites were known for iron working near the Turkish Anatolia plateau the period between 1400 -1200 B.C. It would not be until the age about 700 B.C. that smelting iron would be commonplace.

The Philistines had learned the secret of casting iron by 1200 B.C., and guarded their knowledge jealously. They monopolized the iron trade "There was no smith found throughout all the land of Israel." Deuteronomy 8:7-18 "a land whose stones are iron and from whose hills you may mine copper."

The Greeks used tools and weapons of iron from the 1100th-100th century B.C. There is evidence that they were also counting with the use of decimals. In the 8th century B.C. Homer wrote of Hephaestus melting copper and tin to make Achilles shield. He also wrote "...a blacksmith plunges a screaming great axe blade or adze into cold water, treating it for temper", indicating a knowledge of quenching. In 965 B.C. Solomon, The Great Copper King, imported Phoenicians who were known to be the smelting technicians.

By the second century B.C. China was using coal in large scale iron smelting. China had also learned how to make steel from cast iron. They also used wrought steel extensively. Cast iron was named raw iron, steel great iron, and wrought iron was known as ripe iron. In the classic Huai Nan Tzu, dated 120 B.C., there is a description of decarburization by blowing oxygen over the cast iron. The Han dynasty nationalized all cast iron manufacture in 119 B.C.

Circa 500 B.C. India commenced making production quantities of high quality steel from wrought iron that was called "Wootz". The steel was exported as far as China. In 1722 Reaumur wrote in his "Memoirs on Steel and Iron" on steel from India, "I could find no artisan in Paris who succeeded in forging a tool out of it."

In Periclean times (490-420 B.C.), Orpheus while describing Daedalus's life in Crete credited him with the ability to cast bronze statues by the lost-wax method. The investment casting method is still in use today and the claim was made that it was invented by Professor B. Cellini in 1540.

Paper: Advances in mathematics would not have been possible without the invention of writing, considered to be the world's most important invention. The first manufacture of paper is credited to the Sumerian civilization in the period 3200-2000 B.C. The major invention of paper originated in China in 322 B.C., more than a thousand years before its appearance in the Middle East. Paper from pulp as we manufacture it today was invented in A.D. 105 in China by Ts'ai Lun and was made from bark, fishnet and bamboo. Paper appeared in Spain in 1056, followed by Holland in 1322, England in 1494 and in the U.S. (Pennsylvania) in 1690. Paper from wood pulp was invented by an English man Hugh Burgess in 1852.

The Maya were also making paper from the fibers in fig tree bark and assembling books. The paper was whitened with plaster. Only four books are known to still exist and when the method was first discovered is unknown.

During the Ptolemy dynasty (290 B.C.) the library of Alexandria was built and contained over 700,000 papyrus scrolls. The library was the world's most important learning center of the period.

Machines: The literature that has survived from Ancient Greece and Rome contain many references to mechanical devices that may or may not contain gears. We have few archaeological records as gears were mostly of wood construction and have rotted away. In the 8th century a boring tool was described by Homer (Od.ıx, 384 et seq.); this may well be the first cutting tool, even older than the lathe and the potters' wheel.

"Bored it into the hole: as a shipwright boreth a timber,
Guiding the drill that his men below drive backward and forward,
Pulling the ends of the thong while the point runs round without ceasing."

Figure 1-1 Early Lathe Tools

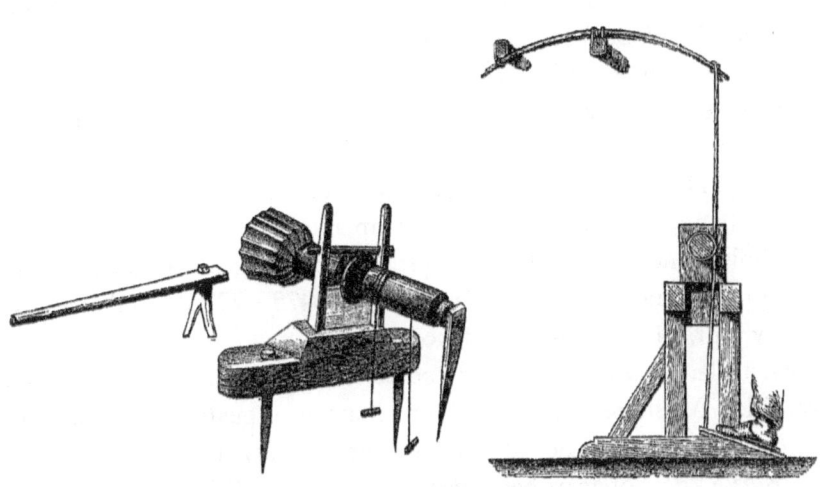

The lathe would be introduced into Europe by the Greeks in the seventh century A.D. Probably the tool was similar to the first illustration in Fig 1-1, both are of the oldest lathe designs; the bow-lathe was still in use in the last century by the Mongol tribe of Kalmucks and the Chinese. The bow-drill is known to have been used by the Egyptians circa 1500 B.C

In China crops were irrigated by using by using a vast system of canals and dams. The Zheng Guo Canal (circa 246 B.C.) was 150 kilometers long irrigating 80,000 hectares. Construction of the canal required advanced technology and machines.

Calendar: Julius Caesar reformed the luni-solar calendar that had required the addition of a month every two years. When the new calendar was introduced, on the first of January, 45 B.C., the old calendar was in error by three months. The Maya calendar which was based on intermeshing the Sun, Moon and Venus with a more accurate "gear ratio" was more precise. Hillel II, the Nasi of the Sanhedrin AD 330-65, worked out a mathematical formula for calculating a calendar that has been used ever since. Previously the Egyptians used a month with three ten day weeks, adding five extra days at the end of the year. Babylonian, Greek, Hebrew and the Chinese used alternate 29 and 30 day months adding an intercalary month every so often. Caesar created the solar Julian calendar that would be used in Europe until its replacement in 1582 when October 4th became October 15th. Pope Gregory X111 is credited with introducing the Gregorian calendar. The calculations were provided by Copernicus, Christophorus Clavius, and the physician Aloysius Lilius The Julian calendar was not adopted by Britain or the U.S. until the 18th century. Britain adopted the Julian calendar on September 2nd 1752, that day instantly changed to September 14th. The League of Nations established a Committee for Calendar Reform in 1923 with zero results. The U.N. recommended a "World Calendar" in 1954 that also went nowhere.

Gears: During this B.C. era gears would necessarily become more in demand. In approximately 4000 B.C. Sumerians used the wheel and gear driven hoists, and by 2,600 B.C. complex differential gears would be in use. During the Chinese period of Warring State (475 - 221 B.C.) the curved tooth cylindrical gear was in evidence.

Aristotle, Greek philosopher and scientist (384-322 B.C.), wrote a book titled Mechanics detailing bronze and iron gears. He defined an object as being hard when it does not yield to penetration through its surface. We know of gears being used on Greek windlasses 300-260 B.C. The Alexandrian Greek, Ctesibius, developed water clocks that utilized racks, spur and bevel gears. He also invented the force pump, water organ, and was Heron's teacher.

In 250 B.C. gears were used to operate hydraulic organs, the first keyboard instruments. Air was blown into a chamber that contained an inverted metal bowl. Increasing air pressure forced the water out of the bowl, raising the level of the water, and forcing the surplus air into a pipe chest above the water cistern. Between strokes on the pump the metal bowl air pressure was kept constant. Some two hundred years later Heron designed a water organ powered with a windmill replacing the former hand pump.

The Byzantium scientist Philo In approximately 230 B.C. wrote a treatise on military engineering of which some fragments remain. He describes a rack and pinion system for raising water. He also wrote on the subjects of elasticity and metal testing. In 200 B.C. we know that oxen powered hoists had gear driven horizontal and vertical shafts.

From the writings of Pliny, Gaius Plinius Secundus the Elder, we know that in this century (82 B.C.) the screw press was developed. This notable event was related to the development of toothed wheels and gear trains. Gears transmitted heavy power in the machinery of the water mills. "Precision" or "Mathematical" ring gearing with exact high ratios have been found and appear to have been in fairly popular usage.

A Gear Chronology

An astronomical computer for the motions of the sun and moon, the Antikythera Mechanism, was made using complex gear trains, including epicyclic systems. This is the earliest surviving mechanism of mathematical gearing. (Fig. 1-2).

Figure 1-2 The Antikythera Mechanism (Circa 82 B.C.)

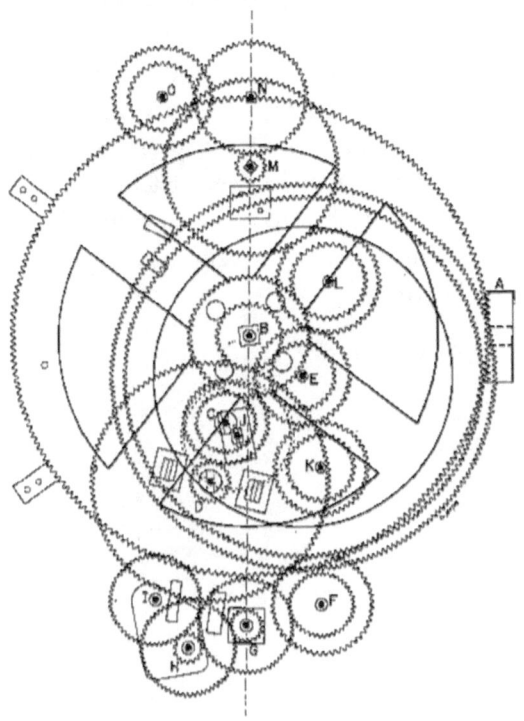

The instrument was discovered in the early 1900's and named for the island off the Greek South Coast where an accurately dated wreck was found. Similar designs built over the next thousand years have been discovered in the Middle East. more than thirty bronze copper and tin alloyed gears were constructed with a 30° pressure angle. Except for the main drive wheel all well preserved teeth are the shape of an equilateral triangle. For ease of manufacture it is believed that the gears were made with an even number of teeth then re-shaped to make the odd number required for the calculations. The input axle came through the casing and turned a crown gear which itself turned a large driving wheel with four spokes. Axles turned two gear trains leading to an epicyclic turntable arrangement.

When considering the development of the gear in all probability one would consider the simplest gear, i.e. the spur gear, would be the earliest and then the helical, bevel, worm and the hyperboidal to be the logical progression.. It would be a mistake to believe that this was in fact how the gear developed. In the Middle East today water is obtained by using a large vertical cog wheel driven in turn by a large lantern pinion. Equally primitive ginning rollers for Indian cotton can be seen in the London Indian Museum that are driven by parallel screw gear wheels. Lantern pinions are made with two wooden discs separated by spacer bars. The cog wheel entering the spaces between the bars. Even today at least one manufacturer, the Nexen Group, promotes a similar roller pinion system. They claim the system has zero backlash, speeds up to 36fps, quieter operation, minimal lubrication, low maintenance with an extended operating life. They improved on the original design by having the spacer bars supported in bearings.

Chapter 2

Standards - Metals - P/M - Mathematics – Machines - Paper – Gears - Mills

Gear Progress to the 13th Century

With the collapse of the Roman Empire it would be many centuries before any significant advances in the use of gearing and its related subjects could be observed. The developments that did occur were mainly in China and the Arab world, especially in the field of mathematics and agriculture. From the 9th through the 16th centuries, Islamic societies from Spain to Oman experienced a "Golden Age" of science, mathematics and technology. Many machines were developed because of the growth in cities and the agricultural need for water. The irrigation techniques and water distribution methods had come from Asia, India and Rome. Following the fifteenth century, techniques developed in the Arab world would be adopted as far away as the Texas and Louisiana sugar fields. From the sixth through the fifteenth century technology moved from the East to the West. Power sources had been animal, human, and water, but in the seventh century the windmill, originating in Persia, became a major power source. The aforementioned power sources continue to be used to the present day. These would be the only power sources available until the second half of the nineteenth century.

In this first millennium A.D. we can clearly see the advances in mechanical and gear technology are driven by the demand for specific machines. The first demands were for the drives needed by the water and wind mills, then clocks. Requiring precision, clocks needed accurate gears. Several gear types were required, needed to be developed, and would in turn contribute to drives for other machines. Crude gears would have continued to be shaped by hammering but for clocks that required something much better. By the 11th century the Arab world would be making clocks with complicated gear trains to be followed by Europe in the 14th century. Lewis Mumford in his 1934 book "Technics and Civilization" wrote the second millennium would have three phases:

ecotechnic (wind, water and wood complex) from 1000 to 1750

paleotechnic (coal, iron and steam) from 1700 to 1900

neotechnic (electricity, hard alloys and lighter materials) in the 20th century.

Animal power was not mentioned but would make an important contribution until the 1950's. Frequently a train of gears was required between the power source and the machine. The drives were of two types, in the older method a large diameter gear was mounted on a vertical shaft above the height of the animal. The gear wheel engaged another gear on a lay shaft that takes the animal-power. Depending on the size of the gear wheel the horse could be harnessed inside or outside the wheel's circumference. With the use of cast iron a low-level version would become possible. A small gear wheel would be encased in a frame from which arises a vertical shaft connected to the horse. The animal had to cross over the drive shaft in this method and this design would persist for many centuries.

In these early years, and in many parts of the world today, obtaining a source of water was critical to survival. A method for raising water from wells was known in Roman times. The word Sàqiya is used for the chain of pots powered by one or two animals walking in a circular path, or on a treadmill and utilizing two gear wheels. Sometimes it was even driven by water power, and it is then known as Na'ura. Schieler provides a detailed study of the machine in Ch. 4, "People of Isfahan", Oxford M.S. 954 He wrote "The crux of these machines is the gear, which has a single function – that of altering the motion from horizontal to vertical. The gear itself consists of a lantern pinion and a large cog-wheel... the lantern pinion is fixed to the upright shaft by means of two sets of spokes, one for the upper rim and one for the lower... the pins of the lantern wheel are slightly conical in shape and are of a soft wood because they are easy to make and the peasant puts in new pins approximately every other year."

A further description of this important machine is in the twelfth century agricultural manual written by the Spanish Muslim Ibn al-Awwām. He indicated an understanding of the fly-wheel principle by suggesting the pot garland elevator wheel should be made heavier. The machine has maintained its importance in Middle East out of way places, its importance can be gathered by the many works on the subject including a full description in the 17-volume encyclopedia written by Ibn Sida, who died in 1066. In the 12th century the Arab world was using a pedestal crossbow. The crossbows were drawn by a windlass that was operated by a rack-and-pinion gear set.

Standards: In the thirty-fifth of the sixty-three clauses of the Magna Carta, dated 1215, King John promised standard measures of capacity, area, and weight. Edward I established the yard as the British standard in 1305. The yard was known as the Iron Ulna. It contained three feet, each foot twelve inches, and an inch was made 1/36th part of a yard. The yard standard that is still in existence was established by Henry VII in 1497. It was superseded in the reign of Elizabeth the First.

They can both be seen in the South Kensington Museum Science Museum, London, England. Standard measurements and the importance of keeping them current gained in importance. In 1742 the Royal Society proposed legalizing the yard standard and this was done by an Act of Parliament in 1824.

Metals: The treatment of metals was well developed in China. They understood, as did the Roman, Pliny the Elder (23-79), that there were variations in different waters used for quenching. Sir John Pettus wrote in 1670 "Pliny saith that the goodness of steel ariseth from the goodness of iron-mine from whence it comes, with the assistance of Waters or Oyls." In 6 A.D. the blacksmith Qiwu Huaiwen used different animal urines and greases to provide a variety of quenching rates.

From the eighth century onwards Styria and Carinthia, provinces of Southern Austria, were making steel, in the main by the cementation process. The steel was sold in countries as far away as England and Turkey.

Theophilus, believed to be the Benedictine monk Roger of Helmarshausen, was a German metalworker and writer. In 1145 he is credited with Europe's first technical manual "De diversis artibus". The manual is divided into three books, book three "The Art of the Metalworker" is twice as large as book one and two combined. Theophilus described the making of brass, casting, forging, and a type of lathe to shape molds. He suggested several quenchants "Tools are also given a harder tempering in the urines of a small red-headed boy than taken in ordinary water." An alternative was to use the urine of goats fed on ferns for three days.

Of thirty 12th century French documents dealing with hammer forges and iron metallurgy twenty five were by Cistercian monks. In 1340 the first blast furnace was built in Liège, Belgium.

Powder Metallurgy: The fifth century legend of Wayland the Smith is an early story of powder metallurgy. Several accounts relate to his making a sword by conventional black-smith methods, then reducing it to small particles, which were mixed with ground wheat. These pellets were then fed to geese and their droppings collected.

The droppings were then reheated and hammered into a sword, and the whole process repeated. It is possible that these, the finest of swords, were benefited in part due to the high ammonia content that nitrided the particles. Other powder metallurgical methods were in use in pre-Columbian times by the Ecuadorian Indians.

Advances in Mathematics: In the year 130 A.D. in China, Hou Han Shu used $\pi = 3.1622$. The Hindu mathematician Aryabhata (A.D. 499) summarized in his "Aryabhatiya" the total mathematical knowledge known at that time. Included was the following statement: "Add 4 to 100, multiply by 8, and add 62,000. The result is approximately the circumference of a circle of which the diameter is 20,000."

$$\pi = 3.1416$$

Liu Hui had reached the same result in China by using a diagrammatic method in 264 A.D, The Chinese mathematician Zu Chongzhi (429-500 A.D.) was the first to calculate pi to seven decimal places (3.1415926). He also made notable contributions to mechanical design and mathematics with his book "Zhui-shu" that would be used for the next five hundred years as a text in Japan, China and Korea.

A significant advance took place prior to the year 300 when the Greek, Diophantus wrote his thirteen books titled "Arithmetica" that were confined to problems in algebra, and the theory of integers. They were the first books to systematically use algebraic symbols.

Dionysius Exiguus (c.500-545) in calculating Easter tables incorrectly arrived at the date AD 1, and the designations B.C. / A.D. Some claim the B.C. / A.D. dating owes its use to the Venerable Bede.

In the third century Pappus of Alexandria wrote a mathematical *"Collection"* that covered a wide range of geometrical problems. At this time they considered three classes of geometric problems, which they named as plane, solid and linear. Plane problems were solved with straight lines and circumferences, solid problems involved one or more conic sections, linear covered those problems that used spirals, quadratrix, conchoids, and the cissoid. Simpler and more useful methods for calculating the surface areas and volumes of solids were formulated in what we now refer to as the two laws of Pappus. The 17th century mathematician Descartes used this work to further develop our understanding of geometry.

A plate from India (circa 595) provided us with the first indication in the use of the decimal system in this part of the world. The system was already known to the Chinese, Greeks and Maya.

The 9th century saw the development of the Indian concept of the sine, (the ratio of the length of the side of a right angle triangle opposite an acute angle to the length of the hypotenuse). Habash al-Hasin introduced tangents to facilitate geometric calculations. Over a period of 150 years the Arabs translated all available Greek books of science and Arabic replaced Greek as the universal language of science. The Arab mathematician al-Khwarizimi (Circa 830), originally known as Mohammed ibn Mūsâ al-Khwârizmî, writing in Baghdad, introduced algebra, algorithms, and Indian-Hindu/Arabic numerals. His writings translated into Latin were influential in introducing Hindu or so called Arabic mathematics to Europeans. His 825 A.D. treatise "The Book of Addition and Subtraction According to the Hindu Calculation" included "...using the nine characters ... capable of expressing any number... tenth figure in the shape of a circle." The Maya and Chinese had used the zero digit a thousand years previously. By corruption of his Latin name algarismus the word algorithm was obtained. From the title of his book "al-jabr" the word algebra was derived. (Arabic jabara, "to restore") Robert of Chester, England, translated al-Khwarizmi's book using the title "Algebra". system (0 thru 9) that we now know as Arabic Numerals. The Mesopotamian monk Sebohkt introduced the Indian numeral to the West. An edict of

1259 A.D. forbade the bankers of Florence from using infidel symbols, and the exclusive use of Roman numerals was still the rule at the Padua University in 1348.

The 11th century physicist, Alhasan ibn al-Haitham, known in the West as Alhazen, by devising the first pin-hole camera introduced the experimental method of proof, insisting that theories had to be verified in practice, a key element that was missing from the less empirical Greek tradition.

Al-Biruni wrote the first book on trigonometry in the 10th century. It became a separate discipline when it was further developed by al -Tusi the 13th century astronomer.

The period 1100 -1250 was an important period of transition and absorption of the more advanced Muslim mathematics and sciences into Europe. Prior to this period ignorance, disorder, and religious isolation had prevented the West from advancing in science and engineering. Many historians call what followed a renaissance period. The core of the civilization in Western Europe was by the middle of the thirteenth century Greco-Arabic-Latin. Importantly the Hindu numerals became known largely through the work of Adelard of Bath, England, (1080-1152). He studied in Antioch, Turkey and discovered the mathematical system of Euclidean geometry. His pupils used both Latin and Hindu numerals. They did not become into wide spread use because of prejudice and a lack of need. In the twelfth century the only European mathematicians were from Spain, namely Abraham bar Hiyya and a Moor Jàbir ibn Aflah. A century later there would still be only four or five mathematicians of note, Fibonacci, Nemorarius, Grosse-teste and Bacon, all using Greek, Arabic or Latin systems.

The Hindu mathematician Bháscara was born in 1114. He wrote, as far as we know, the oldest book dealing comprehensively with fractions. The denominator was written under the numerator without any demarcation line. Arabs later introduced the line as a symbol of division.

A history of mathematics and science "Liber abaci" was written by the Italian Leornado Fibonacci (1202). Fibonnaci was the outstanding mathematician of the middle ages. The treatise contained the first complete account of Hindu numerals. The book helped establish the universal Arabic numeral system. He also originated the Fibonacci Sequence, the first recursive numerical sequence. Each number is equal to the sum of the preceding two, 1, 1, 2, 3, 5, 8, etc. (A sequence in which two or more successive terms can be expressed by a formula). Fibonacci made another major contribution to numerical theory in 1225 with his book on square numbers, "Liber Quadratorum".

Machine Development: The basic essentials of the piston and cylinder steam engine, lacking only the crank-shaft, were invented in China. The details are provided in the book "Description of the Buddhist Temples and Monasteries of Loyang" written about 530 A.D. This water powered machine sifted flour and operated in a reverse mode to later steam engines. The machine wheels were driven by rushing water to power the pistons. A later use of this power was to operate giant bellows for blast furnaces.

In 723 A.D. Buddhist monk and mathematician, I-Hsing, developed a "Water Driven Spherical Bird's- Eye-View Map of the Heavens", with machinery regulating the movements.

In Baghdad approximately 830 A.D. the three brothers, Muhammad, Ahmad and al-Hasan Banū Mūsà, all prominent engineers and scientists, completed their book "Kitabal-Hiyyal (Book of Ingenious Artifices)" which included descriptions of over one hundred machines. Their work "On the Measurement of Plane and Spherical Figures" was unsurpassed until modern times.

The Utrecht Psalter in Leyden University, dated 850 A.D., illustrates the first mechanization of grinding, a hand cranked grinding wheel. Previously grinding was entirely a hand operation.

Note: The Institute for the History of Arab-Islamic Science, Frankfurt, Germany, contains more than 800 replicas of scientific and engineering innovations.

Paper: Paper was to be increasingly important for the recording of technology and mathematics, and thereby education and machine design, particularly in the second millennium. The Chinese used silk and then linen to make the earliest paper. Paper made from pulp was invented by Cal Lun during the Eastern Han dynasty (A.D. 25-220) using rags, hemp, bark, and discarded fishing nets. In 794 A.D. state owned paper mills were established in Baghdad. The oldest surviving paper manuscript was written in Greek on Arab paper.

Leyden University possesses the oldest surviving dated book, "The Book of Linguistic Difficulties in the Traditions of the Prophet". The book was written by Kitab Gharibal-Hadith in 866 A.D.

At the end of the 8th century Harun al-Rashid erected Baghdad's first paper mill, the second in the empire. The first mill was built in Samarkand about 750 A.D. by Chinese engineers captured in the Battle of Talas. The Chinese had kept their process secret since the 2nd century. The mills would proliferate over the Middle East. A key technological breakthrough for the spread of engineering science. For the next two hundred years there would be a massive undertaking to translate Greek, Persian, Syrac and Indian treatises into Arabic.

In 952 A.D. Abu al-Hasan Ahmad ibn Ibrahim al-Uqlidisi, the mathematician known as the *Euclidian*, wrote a treatise to adapt to the new use of ink and paper, altering the Indian/Hindu method of calculation. The use of paper gave greater flexibility to calculations. Previously mathematics had been written on a dust-board with numbers erased and shifted.

In Theophilus's first book *"The Art of the Painter"*, dated 1145, he recommends *"Byzantine Parchment'* which is the first written reference to paper in the West.

Gear Developments: By the year 25 the Romans had started to cast bronze gears. Vitruvius Pollio, Marcus the Roman architect, and an engineer in the service of Augustus, wrote ten books under the title "de Architectura". They are the only Roman architectural treatises known to still exist. There are few ancient books that contain illustrations of machines and this is the most notable. The books included information on mechanical engineering, a number of techniques and machines, worm gears, and a water clock with rack and pinion gearing. This work was completed about 27 A.D. and included the first clear description of a vertical water-wheel powering a flour mill with one pair of toothed elements ratio 5:1. An early authenticated instance of gears used to transmit power. The gears could be adapted to the speed of the stream. In later designs the crown wheel was mounted on the water-wheel shaft, and a lantern wheel mounted on the mill's shaft "… at one end of the axle a toothed drum is fixed. This is placed vertically on its edge and turns with the wheel. Adjoining this larger wheel there is a second toothed wheel placed horizontally by which it is gripped. Thus the teeth of the drum which is on the axle, by driving the teeth of the horizontal drum, cause the grindstone to revolve." The wheel he described was undershot. Typically the efficiency ranged from fifteen to thirty percent, whereas an overshot wheel has efficiency from fifty to seventy percent. Vitruvius mentions several Hellenistic engineers including Ctesibios. Francesco di Georgio (1439-1501) translated Vitruvius's "de Architecura", in Venice Danile Barbaro also translated this work in 1556. Water powered mills were rare in the late Roman Empire, vertical wheels and overshot wheels in the minority, and almost always used only for grinding grain.

Heron of Alexandria wrote three books on mechanics. In book one the gear ratio effect is explained. Book two describes five basic machines the lever, pulley, wedge, screw, and wheel with its axle. The mathematics of a worm is detailed with its pitch, thread spacing, and appropriate profile. Book three provides the practical application of mechanics to machines. Also of interest to gear engineers are Heron's three other books under the title "Metrica", all treatises in pure geometry mensuration.

Between the years 75-100 A.D. Heron invented a hodometer or cyclometer to measure the distance traveled by freight wagons using brass worm gears to provide a high ratio gear train. According to M.R. Cohen and E.I. Drabkin's book "A Source Book in Greek Science" a pin was attached to the carriage wheel hub. The pin engaged the first gear wheel, a bronze disc fitted with eight equally spaced pins. When the carriage wheel made eight revolutions the gear moved one revolution. A worm drove a gear with thirty cogs, using a carriage wheel ten cubits in diameter; a thirty cog revolution measured a journey of 1600 cubits. A practical limitation on the number of gear trains also allowed for a system that could be reset to zero for the next journey.

It has been reported that Vitruvius mounted a four foot diameter wheel to a frame. He concluded that four hundred revolutions of the wheel was equivalent to a mile. The axle of the vehicle was fitted with a four hundred toothed gear and a pin. Each rotation advanced the gear one tooth. After a complete revolution of the gear a pebble would drop into a container. At the journey end the total number of pebbles would provide the mileage travelled. There is no evidence that the mechanism was ever built.

Figure 2-1 Heron's Odometer

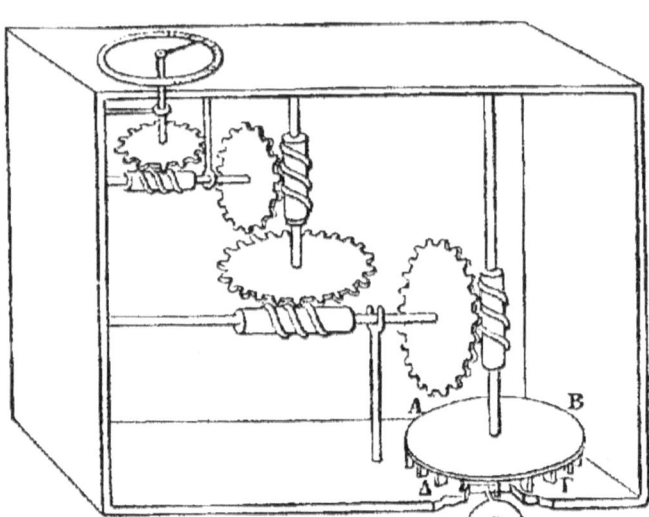

On the Mormons journey to Utah Clayton calculated that a wagon wheel turned 360 times to travel one mile. By tying a red rag to the wheel he calculated the revolutions. Fortunately there was a skilled carpenter Harmon and a mathematician named Pratt. They attached a bar to a gear that moved forward one tooth every quarter mile. Checking the amount the gear advanced provided them with the mileage. The mechanism, they called a roadameter, was totally enclosed to protect from dust and rain. The same principle applies to the modern automobile. A mechanical odometer is connected to a tiny gear that turns with the vehicles motion. One hundred and sixty-nine times registers one tenth of a mile.

Heron made and exhibited in the Serapeum, Alexandria, a rotary type steam engine figure 2-2, known as the "Aeoliphile" Father VerbiHeronest, a Jesuit missionary in Peking, China, used an aeophile in 1630, "with jets of steam playing on a revolving winged wheel geared to the wheels of a car."

Figure 2-2 Aeoliphile

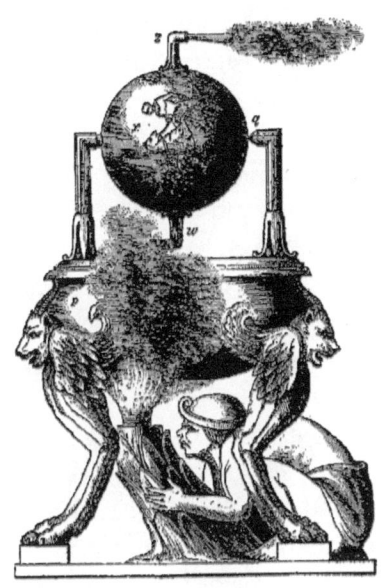

Figure 2-3 Chinese Cart with Geared Speedometer

During the Chinese Jin Dynasty, 947-- 950 knowledge of gearing enabled the construction of a two horse powered cart with an odometer. A decorative figure struck a drum every (li) half kilometer while a similar figure would strike the drum every five kilometer (10 li) as shown in figure 2-3.

Parts of a Byzantine Sundial-calendar dated 480-560 A.D. were obtained by the London Science Museum in 1983. The same size triangular shaped gears are made of an alloy of copper and zinc, and not copper and tin as used in the Antikythera Mechanism, although it was a continuation of that mechanism's design. Similarities in the gear design and layout would reappear in the Arab world around A.D. 1000 providing strong evidence of the influence of Greek technology.

By the end of this millennium gears were in common usage in the Arab world for applications ranging from water clocks to water wheels. In 850 A.D. Arabic treatises were written on the astrolabe, a geared navigational instrument that was used for navigation prior to the invention of the sextant. The one surviving example of Islamic mathematical gearing was made by Muhammad Abī Bakr ar-Rāshidī al-ibarī al-Isfahānī in 1221/2 for an astrolabe. Signed and dated, it is in the History of Science Museum, Oxford, England. The number of teeth in the astrolabe calendar ranged from eight to sixty-four and provided the sun and moon position in the zodiac and the ages of the moon. The earliest surviving gear mechanism from the Latin West were also used in an astrolabe, dated c.1300. This astrolabe was made in France of brass. A ring gear with 180 internal teeth and a number of gears mounted to an arm have survived. The 39 tooth outermost gear engaged inside the ring gear so that when the arm is moved the train of gears also went into motion. This gear in turn engaged a 27 tooth which engaged a 45 tooth engaging a 15 tooth pinion that engaged a central 24 tooth gear. Arabic numerals, introduced into the Latin West in 1202, were used for the numbering scale. The teeth were filed using punched dots and radial lines as a guide. Arabic numbers read right to left so as we write 15 then it would be written as 51.

The vehicle in figure 2-4 was first constructed by the Duke of Chou in China's Honan province, possibly 1030 B.C. and an official history was written in 500 A.D. A similar vehicle had been made by the famous scientist Zhang Heng (78-139) with the probable date of 120 A.D. The machine's importance to the gear enthusiast lies in the fact that it used a train of differential gears similar to a modern automobile. It is believed that the Greeks had co-invented the differential gear about 80 B.C. others believe differentials existed in 2,600 B.C., and this crucial invention for wheeled vehicles would be reinvented more than once in the1800's. Power was transmitted from the wheels through the differential, in reverse of the method used in the modern automobile. The chariot required that parts be constructed with ninety nine percent accuracy. Figure 2-4 is of a working model made by G. Lanchester.

Figure 2-4 Lanchester's Working Model – London Science Museum

In China's Honan province, Su Sung built a forty foot high astronomical clock-tower as shown in figure 2-5. Powered by a waterwheel, a chain drive drove a celestial globe and an armillary one revolution per day. The water wheel also powered a gear train that operated the bells, drums and a large number of miniature figures that indicated the time. Built in 1090 the mechanism ran for thirty six years and then was moved to Peking where it continued to run for several more years.

Figure 2-5 Su Sung's *Cosmic Engine*.

At the beginning of the second millennium the Persian scholar al-Bīrūnī (973-1048) wrote a treatise on calendrical devices similar to what had been made by the Greeks and in the Byzantine Empire some five hundred years earlier. (Figure 2-6). The gearing was not identical but the gears could be rearranged to fit behind the Byzantine sun-dial calendar front plate.

Datable to the first century B.C. the earliest surviving example of complicated gearing is the Antikythera Mechanism (Fig. 1-2), but it was of negligible power. The similar design of the Byzantines and al-Bīrūnī, were followed in importance by al-Murādi's five machines that have great significance in the history of mechanical knowledge. Built in Spain in the 11th century.

Figure 2-6 al-Bīrūnī Description Calendrical Gearing

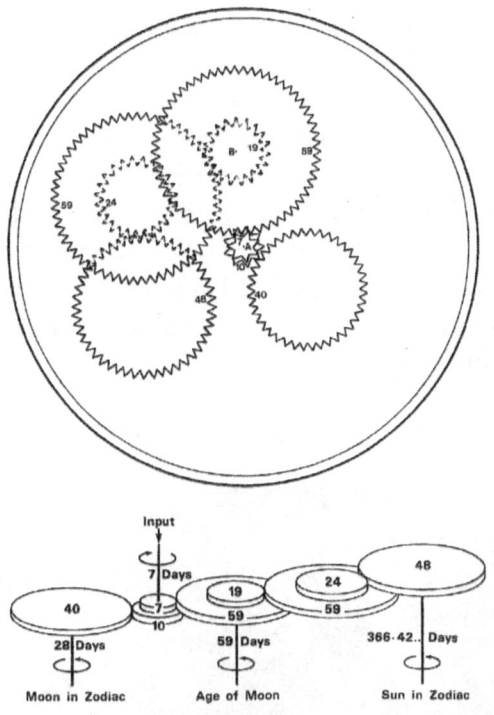

The fifth machine was the first powered water wheel with complex high torque gearing. His treatise on these machines contained for this period four unique gear features, power transmitting, epicyclic, segmental, and made of metal. A drawing of the machine in Biblioteca Medicea Laurenziana, Florence, is badly defaced and it is impossible to discern if the gear train is epicyclic. Smaller gears were made from copper or bronze, the teeth were filed in the shape of equilateral triangles. This tooth shape was easy to produce by hand sawing and filing, and can be seen in Greek work and Giovanni di' Dondi's astronomical clock of 1364.

In 1206 al-Jazari, an Islamic engineer and inventor from Diyar Bakr, Mesopotamia, completed his "Book of Knowledge of Ingenious Mechanical Devices." The book contains colorful, finely detailed illustrations with detailed descriptions for segmental gears, sketches of clocks, irrigation machines, pumps, etc. On his first water-raising machine the fulcrum was a long axle mounted on stanchions near the water source. On this axle a lantern pinion meshed with a vertical segmental gear wheel on another horizontal axle with teeth in one quadrant. At the other end of this second axle a vertical gear meshed with a horizontal gear wheel fixed to a vertical axle. The vertical axle was rotated by a donkey. The segmental gear was well known in the Muslim world, and allowed for regular disengagement of the drive. (The first European appearance of segmental gears was in ' Dondi's clock). The third of five water-raising geared machine designs was built no later than 1254. This machine remained in constant use until 1960 serving a hospital on the River Yazid. It has now been completely restored by the Aleppo University. The design included two right angle gear drives. The Arab world still uses two machines of note, the animal powered saqiya and the water powered na'ura. In the saqiya (figure 2-7) a large vertical cogwheel meshes at right angles with a lantern pinion that may be six feet in diameter.

Fig. 2-7 Saqīya as used in Present Day Pakistan

The late British scholar Donald Hill, and co-author of "Islamic Technology", stated:" It is impossible to over emphasize the importance of al-Jazari's work in the history of engineering." With over fifty machines to his credit he is considered the most versatile of inventors. Al-Jazari is also credited with the first known use of a mechanical crank. His books and the works of other Islamic inventors are preserved in the Topkapi Museum in Istanbul.

Early European machine drawings can be said to have begun with the French architect Villard de Honnecourt (circa 1250). His sketch books provide some of the earliest machine drawings, among which is a celebrated drawing of a self-powered sawmill. Also included were the machines of the time hoists, pumps, and water mills. Honnecourt also wrote on various aspects of geometry using the word "geometry" for the first time. His drawings basic characteristics would remain the machine drawing style for the next five hundred years. He was amongst the first known to use design and construction drawings. His are the only European professional engineering notebooks of the late 14th century.

Water and Wind Powered Mills: Iran's chief invention of this early period was their version of the windmill which was written about in the 7th century. The windmill was of an enclosed horizontal design admitting the wind on one side. The sails were made from wood or bundled reeds. In 1250 the Crusaders returned to Europe with gear driven windmill designs that had originated in Iran. It is thought by some that the horizontal design with the axle at the top of the structure geared to the mill stones was originated in England as a replacement for inefficient tidal mills and a scarcity of water mills that required experience in dam building. Windmills were seen in China in 1219 and it is believed that they were in use in China prior to 600 A.D.

In Europe the vast majority of industrial mills were used for fulling. Fulling is the process in which new cloth was scoured, or cleaned with a detergent to remove the oil and dirt, while being pounded to tighten the weave. The mill axle was required to raise and drop a series of hammers repetitively. The first reference to full size windmills appeared in the writings of al-Istakhri in 951 A.D. The windmills were located at Seistan in the western part of Afghanistan.

A huge expansion in the use of water mills took place in Europe, less so in the Arab world due to lack of water. The Domesday Book recorder 5,624 before it was completed when a year earlier fewer than a hundred were known. The water mill drive figure 2-8 is probably typical of the crudity of most gear drives at that time. A huge further expansion in the number of mills had taken place between 1080 and 1200. In England it was calculated that there was a mill for every 48-50 households. Prior to this time the vertical undershot, overshot and horizontal water driven mills were in use in the Middle East and Europe.

Figure 2-8 Early Water Mill Gearing with Rectangular Teeth

A hunting tooth was in popular use in Chinese mill and boat gearing in 1170. Wind was also extensively used to pump water in Holland, other uses included saw mills, forge hammers, and milling of a variety of products.

CHAPTER 3

Gear Technology from the 14th to the 17th CENTURY

Patents-Science-Metals-Mathematics-Tooth Forms-Clocks-Machine Tools- Tooth Cutting- Applications-Mining-Power- Measurement- Metric.

Introduction: The period known as the Renaissance period is said to have began in 1453. The early leaders were in Florence, patronized by the Medici and this cultural tidal movement that they started moved across Europe to reach its fullness in England after having receded from Italy. The toothed wheel, pin-gear, wheel and pinion were the principal gears in Renaissance technical applications. Mills, machine tools, hoists and similar devices were almost always fitted with one or more of these gear types.

A number of pictorial catalogs appeared soon after 1400 mostly in Southern Germany. Amongst the earliest European illustrated technical manuscripts are those by Mariano di Jacobi detto Taccola who was born in Siena in 1381 and died in 1453, he called himself the "Sienese Archimedes". Leonardo took advantage of his work. Taccola's two books titled "De Ingeneis" and "De Machinis" are considered to be the first in which we can follow a person actually working out technical ideas.

Technical Drawings: Such drawings would make their appearance in the middle ages and would become more common during the Renaissance.

Drawings are dated from the thirteenth century and their first use was for the construction of the Gothic cathedrals. The engineering drawings from this period are in the form of presentation manuscripts. By the middle of the fifteenth century drawings would be in demand for hoists, pumps, mills, raising water, and similar mechanical devices.

Towards the end of the sixteenth century printed books on the subject of Machines by authors such as Dürer (1525). Besson (1584), Ramelli ("the first great engineer who has left a well-established technological oeuvre".(1588), Strada (1618), would become generally available. These books included detailed references and different perspectives that were of assistance to the builder. A prime example of this advancement was Ramelli's geometrically precise drawings illustrating all the essential components of the machinery.

The manuscripts of Guido da Vigevano were one of the earliest in importance to appear. An Italian medical doctor from Pavia and astrologer to the French court, he wrote in 1335 an illustrated manuscript "Texaurus Regis Francae" to inform Phillip VI of France on the necessary equipment and supplies for a proposed crusade. Of interest to the gear engineer was the fact that his dimensionless single drawing per machine and text presupposed that skilled artisans could then make the machine: The ratios and sizes of the gear were entirely dependent on the millwright. Vigevano included a wind powered assault wagon, and about the gearing he wrote "all these matters are the concern of the master millwright and especially the master windmill-wright...a skilled man will easily understand this because I cannot write it more clearly."

The next important development took place with the sketches of Konrad Kyeser, drawn in Bavaria in 1366. The author Bertrand Gille calls Kyeser "the first great engineer who has left a well-established technological oeuvre". He moved from a flat style to three dimensioned drawings. Explaining technology to the non-technical by means of drawings would become common practice throughout the 15th century.

In 1496 the Bavarian gun-maker Philipp Mönch composed the "Buch der stryt". In this detailed document the drive mechanisms were shown separately and enlarged on almost every sheet. It is of particular interest that Mönch favored the selection of gear racks and worm gear drives. Some racks were semi-circular and the worm gearing appears to be quite intricate. Thread cutting lathes were used to cut the worms.

Surprisingly, from approximately 1450 to 1750 machine designers all used a similar graphic presentation with each machine having one single aspect drawing.. By 1750 orthographic projections would be coming into use. In Florence, Italy, in 1563 one of the first technical schools was founded, the Accademia del disegno. Part of the curriculum of these schools was the teaching of making and reading technical drawings, as can be discerned from this school's title. (disegno - drawing). These significant industry changes involved a high degree of skill, and provided scientific advancement in practically every engineering branch except electricity and steam power.

In 1558 a Nuremberg owner of several mines, technician and inventor named Berthold Holzschuler wrote a holographic will for his eldest son. The will only contained a bundle of sketches of what Holzschuler believed to be unique driving methods for vehicles and mills. His intention was to have his son lease these sketches to monarchs, cities etc and become rich in consequence. In each sketch was a logical sequence, displaying mechanisms and components, with an accurate depiction of gears and transmissions. The most interesting last sheet depicted all the details of the drives for a mill with sixteen millstones. The most astonishing fact was that all drawings were true to a scale and drawn using only a ruler and a pair of compasses.

Salomon de Caus in his "La perspective avec la raison des ombres et miroirs", written in 1612, had sketches that combined plans and elevations, or horizontal and vertical orthographic projections.

Patents: Filippo Brunelleschi (1377 – 1446) pioneered patent protection for inventors. He received from the Republic of Venice the very first patent ever awarded. A system of patents was introduced in Vienna in 1474. The system spread and was introduced into England in 1552. In England the system was so abused it led to the English Statute of Monopolies of 1623 which corrected some of the flaws and granted exclusive rights under letters patent for no longer than twenty-one years to the first l inventor of a new technique. It became the model for most future patent laws. Massachusetts "Body of Liberties" (1641) included "no monopolies but...new inventions... for short time only." Under this authority Joseph Jenks was granted a patent in 1655. In1691 South Carolina passed the first American patent law "for the better encouragement of the making of engines for the propagating the staples of this colony." The U.S. Constitution written in 1787 gave authorization to Congress in Article 1, Section8 "To promote the progress of science and useful arts by securing for limited times to authors and inventors the exclusive rights to their respective writings and discoveries." Although it did not immediately function in the face of state charters, it was a unifying law. Thomas Jefferson established the U.S. patent office in 1836, and gave the responsibility for the examination of the applications to Dr. William Thornton, previously the patents had been personally examined by Jefferson.

In Britain the Crown granted patents and copyright as a royal prerogative, in the U.S. the authority was granted to the executive branch of the new government.

Between 1849 and 1895 over 100 patents would be issued for the cutting of gear teeth. Between the years 1867 and 1930 the U.S. patent office issued 2,344 patents for gear tooth cutting mechanisms. By 1900 U.S. patent production was the highest in the world. Amongst the States, in proportion to the number of people, Connecticut produced the most patents. Edison with 727 U.S. patents was the most prolific inventor.

Chart 3-1 Patent Growth

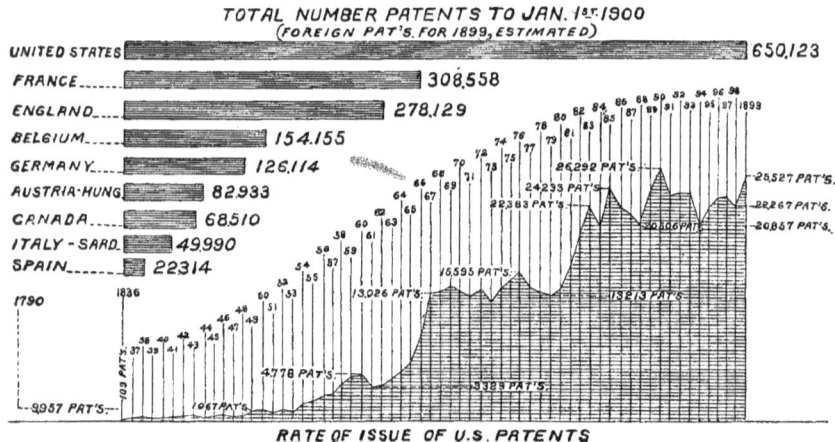

It will later be seen that patents seriously delayed the wide use of steam engines, electric motors, automobiles and other inventions for periods of up to twenty years. The Universal Copyright Law Convention in Paris was accepted by the U.S. in 1974.

Scientific Institutions: During the period leading up to the 18th Century a rapid development in mathematics and the natural sciences would occur. Earlier the universities had been controlled by the church and what slow advances in science and engineering had been made came mainly from the monasteries and cathedrals. Learned men started a movement to create scientific institutions. The first such institution was the Accademia Secretorum Naturae formed in Naples in 1560 and later suppressed by the Inquisition. In 1603 the oldest existing scientific society Accademia dei Lincei was founded in Rome followed by the founding of Accademia del Cimento in Florence. The New Philosophy or Experimental Philosophy was discussed by the London learned in 1645 and would lead to the formation of the Royal Society in 1662. The French Académie Royale des Sciences started with informal meetings of the scientific community leading to its formation in 1666. The Russian Academy of Sciences was formed in 1725, and the Berlin Academy of Sciences in1770. All these organizations published their findings which assisted in rapidly advancing mechanical knowledge.

Progress in Metals: Iron had been a major resource in Britain since 500 B.C. During the middle ages ever increasing quantities were being mined and smelted to supply the demand. The iron was being used for ships, military, agriculture and domestic tools and utensils. A major product improvement took place in 1496 with the invention of the blast furnace in the English county of Sussex. By 1700 the technique had become widespread. In 1625 Lord Dudley William Oughtred, born in Eton College and educated at Cambridge, was issued British patent #18 for improved iron production. "The mistery, arte, way and meanes of melting iron owre, and of making the same into cast workes or barrs with seacoales or pittcoales in furnaces with bellows of as good condition as hath bene heretofore made of charcoale." This is considered to be the start of modern iron working. Slitting mills to handle iron bars were in use in 1649 at Dartford Bridge, U.K.. The first American iron works was set up at Saugus, Massachusetts in 1646.

"De la pirotechnica" was published in 1540, it was written by the Siena born Vannoccio Biringuccio. This was the first practical book on metallurgy, and the production of metals. He credits di Giorgio as being the first inventor of mines. Biringuccio's method for boring gun barrels: "I had a square piece of steel welded, with all four corners true and sharp, and well tempered, so that when I put it in the mouth of the gun and rotated made it exactly round." Biringuccio also bored two cannons simultaneously by having a toothed wheel on one axle driving a rundle wheel on the second borer's axle.

In England, in 1568 near Tintern Abbey, the first brass was made by alloying copper with zinc.

Also in England, the Cementation Process, the method using coal to convert bar iron into steel, was patented by Ellyott and Meysey in 1614. The patent also banned the importation of steel. Five years later the English ironmaster Dud Dudley, illegitimate son of the 5th Baron Dudley, attempted with limited success the smelting of iron with coal. Dudley wrote on the subject of iron in "Metallum Martis" in 1665.

The Netherlands were the first to think of looking for coal underground. Coal began to rival charcoal in the making of iron, thereby providing increased carbon content and higher furnace temperatures. Early in the 17th century the Englishman Hugh Platt discovered how to heat coal and obtain coke which was a much more efficient fuel.

In Germany the Stuckofen furnace was a great improvement in making iron. The former Catalan furnace of Roman origin had been three or four feet high,. The furnaces would continue to be enlarged so that by the end of the seventeenth century they would exceed thirty feet in height and be in widespread use throughout Europe. The Belgians made further improvements by using a waterwheel to power the large bellows. The great quantities of air forced in by the bellows was considered "The greatest technical achievement" of the period.

Figure 3-1 German Stuckofen Furnace

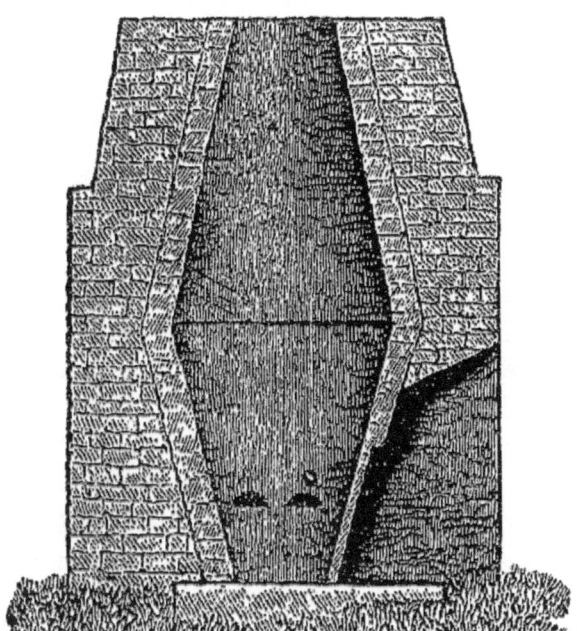

Professor Benvenuto Cellini, Italian goldsmith and sculptor has been credited with inventing the investment or lost-wax casting method in 1540. Although it is believed the method was used by the Greeks during the Periclean period. Casting techniques were further advanced in Holland in 1672 by using loam molds.

In the 17th century, the three volumes of "Exploitation of the Works of Nature" were written by the Chinese scientist Song Yingxing, describing copper and iron smelting. He also describes quenching in oil to provide a gentler quench, "since the strength of steel lies in the quenching." A further notation referred to barbarians quenching in "dison", the urine of the earth, not available in China, it was probably petroleum based.

Giambattista della Porta in his books "Natural Magic" circa 1600 recognized the different tempers for steel based on the color and described methods to achieve them. "...take the chest out from the coals with iron pinchers, and plunge the files into very cold water, and so they will become extremely hard. This is the usual temper for files... If you quench red-hot iron in distilled vinegar, it will grow hard. The same will happen if you do it into distilled urine, by reason of the salt. If you temper it with what in the month of May is found on vetches leaves, it will grow most hard. For what is collected among them is salt..." The cause of quench cracking was also understood. "It is quenched in oil, and grows hard, because it is tender and subtle, For should it be quenched in water, it would be wrested and broken." Porta made a significant contribution with his observation that tempering had a critical range based on the heated steel's color.

Galileo's famous book "Two New Sciences" was the first publication on the strength of materials. Galileo considers the bar's strength "its' absolute resistance to fracture" and that the strength is proportional to the cross-sectional area.

Mathematics: Ghiyath al- Kashi wrote an important mathematical treatise in 1427 called "Risala al- Muhitiya (Treatise on the Circumference)". It included the value of pi to seventeen decimal places. Symbol ϖ was first used by William Jones in 1706.

In 1492 Francisco Bel Pellos's "Compenia de la abaca" introduced a dot to represent division by ten, the precursor of the decimal point. Our present method of writing decimals would not appear until the eighteenth century. One of the first books to be printed in Europe was an impressive tome on mathematics written by the Franciscan Luca Pacioli in 1494. "Summa de Artithmetica, Geometria, Proportioni et Proportionalita" contained all the known arithmetic, algebra, geometry and trigonometry. The German mathematician Christoff Rudolff, who died in Vienna in 1545, introduced "Die coss" in 1525, the variable that we denote as x, the modern symbol for the square root √, and one of the first books to use decimal fractions.

Albrecht Dürer, born in Nuremberg, Germany, re-discovered epicyclic curves in 1497 and also wrote on measurement. The first of his treatises written in 1525 "Underweysung der messung" was for artisans "...who base their art on the correctness of the drawing." The book included proofs on the construction of complicated curves. He invented a highly developed drawing using a combined views technique and also wrote a treatise on geometry.

Figure 3-2 Dürer's Wood-cut of Emperor Maximilian's Coach

An engraver in metal and wood he ranks higher than the painters of his time. Dürer is also considered to be the inventor of etching having produced several plates with all the lines by acid. He was also the foremost artist in formulating *"descriptive geometry"*. Dürer's drawings were examples of combining workshop techniques and theoretical knowledge. His wood cuts and engravings made him famous. His student's wood cut of Emperor Maximilian the First's coach as shown in Figure 3-2 has two rows of wheels driven by a single dual worm gear shaft. The large worm wheels are concentric with the coach wheels.

In Germany, in 1544, the theory of logarithms and exponents was published in "Arithmetica Integra", the work of Michael Stiffel. He brought into general use the symbols =, +, - first occurred in Johannes Widman's "Arithmetic" published in 1489. An Antwerp work by G.V. Hoecke published in 1514 also used these symbols. By the beginning of the 17th century they had become the widely accepted mathematical symbols. The first English book to contain these signs was "Whetstone of Witte" written by Robert Recorde in 1557.

The Italian Niccolò Tartaglia (1500-57), taught mathematics and was praised for his design ability, but also for evaluating designs. He could analyze and define in mathematics the design's capabilities. In 1557 he wrote "Nova Scientia" placing the subject of artillery into mathematical science. Tartaglia was probably the most prominent theorizing engineer of his time. Di Giorgio and da Vinci both copied and studied his work and learned his method for depicting machinery.

The Flemish engineer and mathematician Simon Stevin, Bruges, Belgium, (also known as Stevenus {1548-1620}), published his book "La disme", introducing the use of decimals. Almost immediately they were generally adopted. The most important work on mechanics to this time was carried out by Stevin. He made major advances in the study of liquids, pulleys, the inclined plane, and discovered the triangle of forces. His work on statics provided mathematical proof on the laws of the lever and the inclined plane. An account of his methods and discoveries in Mechanics is in his "De Beghinselen der Weeghconst" published in 1586. Stevin describes a tread wheel driven crane and the advantages of the crank.

François Viète, the most brilliant mathematician of his time, wrote in 1591 the earliest work on symbolic algebra "In Artem Analyticum Isagoge". Viète made important contributions to all the mathematical subjects particularly trigonometry and geometry introducing several new terms that included negative and coefficient. Spain accused him of being in league with the devil when he broke their 500 cipher code for Henry 1V and calculated the value of pi as an infinite product to the tenth place.

The professor of mathematics at Heidelberg had published "Trigonometry" in 1599 followed by tables in 1608 and 1612 in which he introduced the decimal notation for fractions. The notation was also used in the English translation of Napier's "Descriptio", and Henry Brigg's logarithmic and trigonometric tables in 1624. Afterwards the notation was firmly established. (The notation varies with different countries e.g. fraction ½, in U.S. 0.5, in Germany 0, 5 and in Britain 0·5).

Logarithms can be said have been invented by the Scottish mathematician, John Napier, Baron of Merchiston. He gave a summary of his results to the Dane, and greatest pre-telescope astronomer, Tycho Brahe as early as 1584. The first public announcement came in 1614 when they were introduced as a computational tool in "Mirifici Logarithmorum Canonis Desciptio". One of mathematics' curiosities is that Napier constructed logarithms before the use of exponents. Napier made use of the decimal point, and the base e. In his 1615 book "Rabdologiae" Napier described the calculating apparatus, known as Napier's Bones. The "Bones" became the basis for the first mechanical calculator. (See section Calculations and

Measurement). Logarithms were also independently discovered by the Swiss clockmaker Jost Bűrgi (Justus Byrgius) who independent of Napier published a table of antilogarithms in 1624.

The Savilian chairs in mathematics and astronomy at Oxford were founded by the scholar and tutor of Queen Elizabeth the first, Sir Henry Savile. Henry T.R. Briggs was the first professor of geometry at Gresham College, London, and in 1619, the first Savilian professor of geometry at Oxford. In 1616 and 1617 he visited Napier and with his agreement proposed the use of the base 10 for logarithms in place of that used by Napier. Briggsian or common logarithms – logarithms to the base ten, were published in 1624 with the agreement of Napier. Briggs provided logarithmic and trigonometric tables to fourteen decimal places. He also introduced decimals as an operative tool, where previously Stevenus and Napier used decimals only in a statement of results.

The English mathematician at Gresham College, Edmund Gunter, had provided a list of logarithms to seven decimal places in 1620, and the sines and tangents of angles in the first quadrant. In 1624 he introduced a Line of Numbers, which was the basis for the slide rule invented by Oughtred in 1632. Known as Gunter's Scale the two foot rule had scales of chords, tangents, and logarithmic lines for solving navigation problems. Gunter introduced the words cosine and cotangent Without sliding parts it was in truth a scale described in his "Canon Triangulorum". Lord Oughtred made two of Gunter's scales slide one with the other while keeping them together by hand, thus inventing the straight slide rule. Oughtred also wrote extensively on mathematics, notably "Clavis Mathematica" in 1632. This algebra and arithmetic textbook introduced many new symbols including multiplication and proportion.

The founder of modern analytical geometry, the French mathematician Renè Descartes, connected the previously unrelated fields of geometry and algebra in his 1637 treatise "La Géométrie". The treatise first appeared as one of the appendices in his "Discours de la Méthode". The book is divided into three books, the first two on analytical geometry, while the third deals mainly with the theory of equations. The book has been described as "the greatest single step ever made in the progress of the exact science" making modern geometry possible. His method of representing functions graphically was one of the most important advances in mathematics and tooth geometry. Descartes was also the first to systematically classify curves and show algebraic solutions of geometric curves. The book also made an important contribution to the theory of equations. Even with these accomplishments the mathematical solutions to the dynamics of motion were a mystery to him.

The English mathematician Dr. John Wallis's "Arithmetica Infinitorum" was published in 1655 and in which an appreciation of the meaning and use of negative and fractional exponents appeared. He provided clues to calculus and the binomial theorem, and introduced the symbol ∞ for infinity. On November 26th, 1668 his paper "A Summary Account of the General Laws of Motion" was presented to the Royal Society. In 1685 he made another important contribution to mathematics with his book on algebra, and also edited some of the writings of the Greek mathematicians.

The Frenchman Salomon de Caus wrote a treatise on violent forces in 1615 "Les raisons des forces mouvantes" in which is included description of a rolling mill. He also described a machine that used the expansive power of steam to raise water, anticipating the steam engine.

Sir Isaac Newton, English scientist and mathematician, wrote on his first discovery of fluxions in 1665, an early form of differential calculus. In 1673 he advanced Galileo's work with his Laws of Motion.

Gottfried Wilhelm Leibniz, German philosopher, lawyer, mathematician and a principal inventor of mathematical symbols was considered as the last universal genius having studied the whole known field of knowledge including history, theology, Chinese philosophy, linguistics, biology, geology, mathematics, diplomacy and inventing.

Leibniz Cycloid Equation:

$$y = \sqrt{2x - x^2} + \int \frac{dx}{\sqrt{2x - x^2}}.$$

Independent of Newton Leibniz published his system of differential calculus in 1684. The controversy, whether he or Newton was the inventor of infinitesimal calculus was never really settled. Newton's approach was primarily kinematical while the Leibniz method was geometrical. He was one of the first after Pascal to develop a computing machine. In a subsequent paper in 1686 in a formula for the cycloid he introduced the ∫ symbol.

Jean or Johannes Bernoulli, brother of the equally famous Swiss Jacques, was considered the greatest mathematician of his time, they both joined Leibniz in 1687 and using his methods demonstrated that the cycloid is the curve of quickest descent. In 1690 he had applied Leibniz's differential calculus to a problem in geometry, and was the first to use the term integral. He investigated the infinite series, the cycloid and transcendental curves. A logarithmic spiral is carved on his tombstone in Basle cathedral. In 1696 the first text book on infinitesimal calculus was published by the Swiss Guillaume De L'Hopital, based on the lectures of his teacher Johannes Bernolli.

Mechanical analyses had been restricted in the sixteenth century due to primarily to a lack of understanding on the subject of friction. Important friction experiments by Guillaume Amontons took place in France and the results were published by the Royal Académie in a paper titled "De la resistance caus'ee dans less machines" in 1699. The paper included the basic laws of friction.

The force of friction is directly proportional to the applied load.

The force of friction is independent of the apparent area of contact.

Leupold in his first volume of "Theatrum" confirmed Amonton's conclusions. His four points for reducing friction were to run the machine fast, lightly loaded, fewest parts moving in bearings, and all parts are hard, smooth, round and always well greased. In the first half of the eighteenth century experiments to find the coefficient of friction were carried out.

Within five years these basic laws were also confirmed by the French engineer and mathematician Professor Phillippe de la Hire, and entered into mechanics by A. Parent.

Development of Tooth Forms: China had many technological firsts, but few appeared after the 4th century A..D. The Chinese winding mill and the similar Egyptian water mill, the Sakkiah, used gear teeth that were merely rough blocks with a large amount of play between the teeth. During and after the Middle-Ages these tooth gaps would continue to close-up, and a search began for an improved tooth profile.

Fig. 3-3 Early Chinese Gears

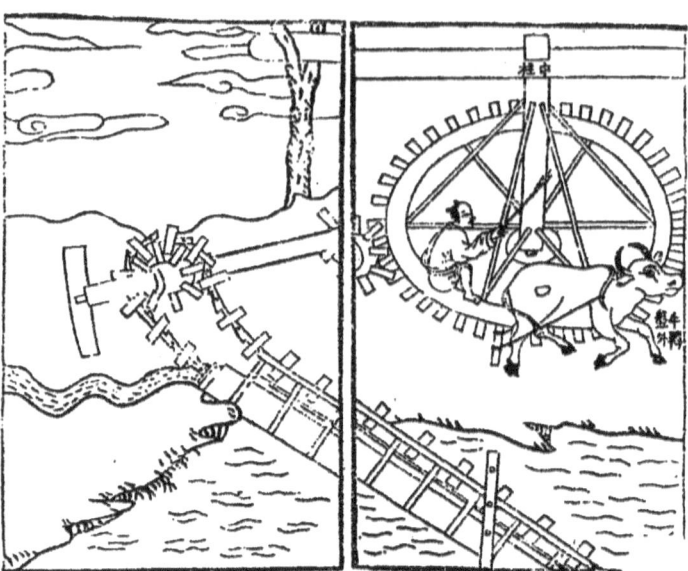

Equipment was very basic and a rare gear illustration as in Figure 3-3 shows round bars were inserted in place of teeth. As the drawing indicates, gears were no further advanced than 1500-2000 years earlier.

Sung Ying-Hsing's 1637 book *"The Creations of Nature and Man"* was published and translated in 1966 under the title *"Chinese Technology in the Seventeenth Century"*. Only water, animal or human power was used in combination with crude tools. Gears were wooden and simple. This classic work provided a history of Chinese technology to the seventeenth century.

Leonardo da Vinci, Italian painter, architect, and engineer made a detailed study of gears. In the early 1490's his notes on machines and mechanics were compiled into his *"Madrid Codex 1"*. It is believed that da Vinci personally bound the two volumes. Some of Georgio's text was copied and included in his Codex. He had met with di Giorgio a few years before when they both consulted on the building of the Pavia cathedral in the1490's. An important difference between them was that di Georgio dealt with whole machines, focusing on certain parts such as the gears, while da Vinci frequently dealt only with the machine's elements. Da Vinci also wrote *"Codice sul Volo degli Ucellie Varie Altre Materia"* showing a highly developed pair of right angle bevel gears (1505). He studied tooth profiles and accurately classified the different power of the teeth. His illustrations included both single and double throated forms of the worm (Fig. 3-4), bevel, spiral and irregular shaped gears.

On the concave worm he wrote: *"This screw acts upon the wheel with equal and continuous force because it is always touched by four teeth."* Writing in reference to worms driving a toothed wheel he wrote "*if only one tooth is engaged and it sheared it would cause great damage.*" Da Vinci recommended helical threads since they prevent the wheels inverse rotation. In his study of tooth profiles he accurately classified the varying power of the motion, when the contact was at the base, middle, or tooth tips. His conclusion was that maximum power was obtained with tip-to-tip combination. *"The top of the tooth must never touch the bottom of the tooth space until the tooth arrives at the point of the central line."*

Figure 3-4 Leonardo da Vinci's Single and Double-Throated Worm Gears

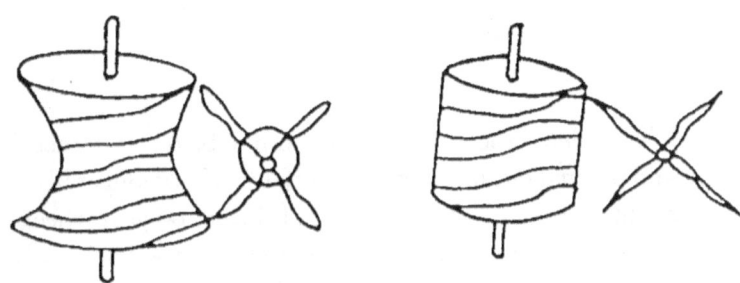

The most commonly used teeth of the time were square shaped, and about these he wrote *"The distance from the middle of the bottom of one tooth space to the other shall equal the length of the tooth."*

Figure 3-5 Laying Out Cogwheel Teeth

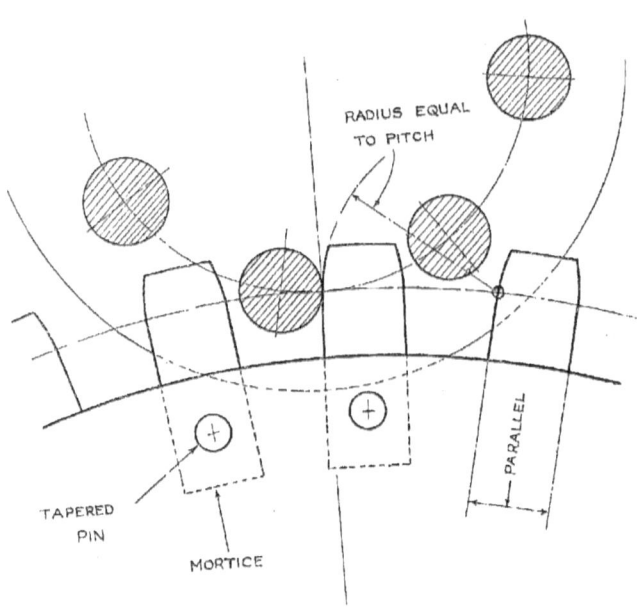

Da Vinci considered the most durable tooth to be the asymmetrical or peg tooth, and pointed out that a helical tooth would last longer because of their larger contact area. At the beginning of the period gear wheels were built in the same manner as heavy wagon wheels i.e. wooden boards were laid in alternate layers transverse to those above and below. Hardwood teeth were mortised into the rim, or near the edge, so as to simplify their replacement when worn. This method has persisted well into the 21st century. The technicalities of gearing were evident.

Origin of Involute and Cycloidal Tooth Forms: Nicholas of Cusa, German philosopher, scientist and churchman was the first to study the cycloid curve in 1451. The Italian Jerome Cardan (Girolamo Cardano) studied the geometry of gear teeth using the limited mathematics available. In Basel in 1557 he provided the first empirical study of gear mathematics that is in print, "*De rerum varietat*". He concluded only two tooth forms were practical for gears, the *epicycloid* and the *involute*. Cardano is also known for his 1545 treatise on algebra, "*Ars magna*", included the formulae for solving cubic and quartic equations. Accused of plagiarizing this solution by NiccoloTartaglia, the real credit should go to Scipione da Ferro, even so it is known as Cardan's formula.

The origin of the involute is not entirely clear. Professor Willis credits Euler (1760) with being the first to suggest an involute gear tooth. However, the involute tooth is also credited to Huygens in 1657, Roemer in 1674. Most credit de la Hire because he published in Paris a complete description and explanation as to the involute's use in 1694. The first individual to realize the involute's practical advantages should be credited to John Hawkins.

Galileo (1564-1647) Italian astronomer, mathematician and university professor gave us the Laws of Uniform Accelerated Motion. Galileo also studied gear meshing conditions, derived the equations, and determined the cycloidal and the tooth root fillet trochoid form.

The Dutch physicist, Christian Huygens (1629-93), invented a tooth shape for crown wheels, and did the first scientific tooth form investigation. The believed superiority of epicycloidal teeth was researched and demonstrated by Olaus Roemer and Huygens in 1674-75. Huygens developed the cycloidal tooth form, and clarified its mathematics in proving the theorem that the cycloid is the tautochrone. i.e., the curve on which a body starting from rest under gravity will reach the lowest point in the same time from whatever point. The "Cycloid" was developed by Huygens during his work on the pendulum which started in December, 1659, and the curve can be clearly observed in his sketches. He was involved in a debate over the curve and was well aware of the properties of its tangent. Huygens is credited with understanding the necessary interplay between theory and practice. He developed a clockmakers lathe in 1657 to cut accurate pendulum clock gears. This accuracy reduced time errors to about ten seconds a day. He also built a vertical lathe with a universal joint for grinding lenses. In 1657 Huygens designed a pendulum clock and continued research until his death. This research resulted in finding the basic fundamentals in the dynamics of moving and rigid bodies.

The Danish astronomer Olaus Röemer was the first to measure the speed of light. More importantly, from our perspective, he also suggested using cycloid and involute gear tooth forms also detailing the advantages of the epicycloidal form in 1674. His work was entirely independent of Girard Desargues the French mathematician who also founded projective methods in geometry with Blaise Pascal. ("Prattique du trait" 1643)

In a memo written to a pupil about 1676 he relates how he has invented a micrometer in 1672.

French mathematician, physicist and child prodigy Pascal, whose father would not permit him to study a subject until he could master it, at eleven years of age he had worked out the first twenty-three propositions of Euclid in secret. At age sixteen he wrote an essay on conic sections. In 1658 Pascal had abandoned mathematics for religion, to relieve the pain of tooth ache, he returned to mathematics briefly and worked out the many properties of the cycloid, making the tooth ache famous. His studies of the cycloid created a controversy known as the Dettonville Problems. While working on the sine function he came close to discovering calculus. Pascal's last mathematical work in1669 was on the area of the cycloid and became the basis for integral calculus. In the same year Sir Christopher Wren had developed a formula for the length of the cycloid.

In 1690 de la Hire developed the geometric principles of gear design, and how to design gear teeth mathematically and systematically. The first mathematical treatise on gear design was by Hire in 1694. Two years later it was translated by Mandy and published in London. Hire reached the conclusion that the involute was the best gear form but it would be 200 years before it gained wide acceptance. He provided a complete description of the involute tooth form and an explanation of its use in Paris about 1694. Tooth surfaces were to roll on one another, and not rub. De la Hire concluded there was a need to maintain uniform pressure and motion. He demonstrated that it was possible to produce the face and flank of any number of gears with the same describing circle. When a tooth is formed by an exterior epicycloid described by any generating circle, the tooth of the mating gear will be part of the interior epicycloid described by the same generating circle. The term epicycloidal in its ordinary use applied only to teeth of which both faces and flanks having describing curves that are circles. Hire made the first systematic application of the epicycloid tooth form in his treatise "Traité des Epicylcloides". He believed that the involute curve was the most ideal tooth form, but his claim to be the first to apply this tooth form cannot be substantiated, and the credit should go to Röemer. De la Hire demonstrated that the direction of motion could be changed by toothed wheels, for which <u>he has been credited with the invention of the bevel gear.</u>

In 1706 de la Hire wrote "It is always possible to find a curve which, by revolving upon a given base-curve, shall generate by some describing point, in the manner of a trochoid, a second given curve; provided that the normals from all points of the second curve meet the first." Professor

Robert Willis pointed out that de la Hire had erred in placing the conical lantern wheels the wrong way, the apex pointing away from, instead of coinciding with the intersection of the axes

Leonhard Euler, a Swiss mathematician who had studied under the Bernoullis, solved the rules for conjugate action in 1751. Having established the basic design criteria he is known as the Father of Involute Gearing. He published several mathematical papers that precisely set out the conditions that must be fulfilled for gear teeth to operate satisfactorily. He showed that either the cycloid or involute profiles would meet the conditions. His work also included a theoretical method on how to produce these forms. Unfortunately his work was too mathematically theoretical to be of practical use to the clockmakers and millwrights. The Swiss ten-franc note carried his picture with a drawing of the involute curve. Euler understood the principle of the common tangent. He also showed how to determine the effect of the tooth on tooth action assuming a pressure angle of 30°. He continued his mathematical work though totally blind, and published over 800 books on every aspect of pure and applied mathematics. His treatises on differential calculus and algebra were used for over a century. Euler published two papers on friction at the Berlin Academy of Sciences. Introduced the symbol μ for F/P, and considered the difference between static and kinetic friction.

In Germany, Abraham Gotthelf Kaestner devised a simple practical method for calculating involute and epicycloidal tooth forms in 1781. This is considered the first step in making the theoretical work practical. Because it was written in Latin it was unusable by the gear makers of the day. His minimum acceptable pressure angle was 15°. Kaestner studied rack teeth and the required length of teeth with the epicycloidal tooth form. He provided a practical way of applying and describing the involute.

Helical Gear: Dr. Robert Hooke, was one of the most brilliant scientists of the age. Hooke was appointed the first curator of experiments at the newly formed Royal Society in 1662 where he held discussions on rolling friction, the wear of bearing materials, and the need for adequate lubrication. Hooke worked with John Wilkins on flying machines, John Willis on chemical research, and John Boyle on air pumps. He is famous for formulating Hooke's Law, and anticipated the steam engine, and Newton's laws on gravitation. Hooke devised the universal joint in 1675 and is said to have invented the microscope.

Amongst Hooke's numerous inventions of most interest to the gear student is his development of the stepped wheel - the predecessor of the helical gear. He cut a number of plates transversely and rearranged them until the tooth of each overlapped the preceding one by the same amount. When there is no sliding between the plates and the teeth are formed by twisting the combination was still known in 1899 as Hooke's Spiral Gearing.

Fig. 3-6 Hooke's Spiral Gearing

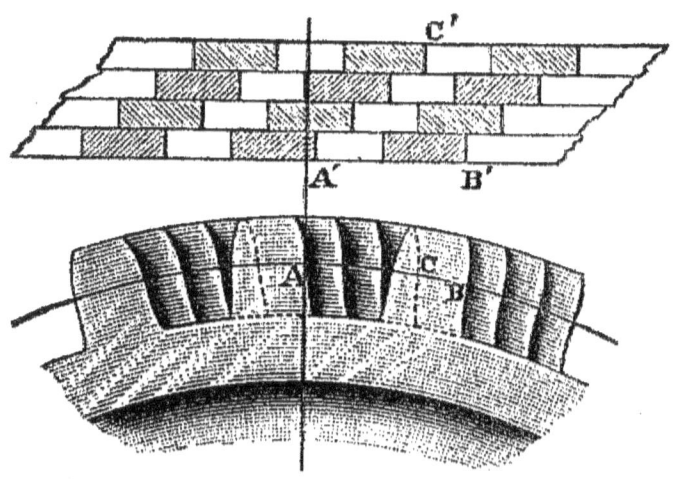

Clock Gearing: In the movement of the West into the technological age the invention of mechanical weight-driven clocks in the second quarter of the 14th century is of major importance. Historical writers such as Donald Hill wrote "one of the main foundations for the development of machine technology..." and D.S.L. Cardwell describes these clocks as "perhaps the greatest single human invention since the wheel." The middle ages were marked by an increased demand for clock and mill gears. No principal city could exist without mills and clock towers. By 1500 few towns in Europe would be without a tower clock. This was a phenomena confined to Europe, and the demand created an increased interest in precision gears. By the 16th century the working parts were accurate and finished to a high level of perfection. Machine tool methods would replace the hand crafted mechanisms at the beginning of the 18th century. Many clocks were highly developed with dials for minutes and seconds, especially those made in Germany. They were accurate enough for astronomical observation which had created the initial demand, they required very large gears. The essential elements of the verge (balance wheel spindle with vertical escapement) and foliot having the balance weight where the crown wheel perpendicular triangular teeth are set, and also engaged at the top and bottom. The arrangement is considered to be one of the most inventive solutions to a mechanical problem ever devised.

The English Abbott of Saint Albans, Richard of Wallingford, reinvented the concept of epicyclic gearing in 1330. His father had been a smith and Richard developed a keen interest in mechanical inventions. He also introduced new methods of calculation that were the beginning of Western trigonometry and built a famous clock in 1320. The earliest Western text reference to the escapement is described in his "Tractus horologii astronomici". The device differs from the verge and foliot and utilized two spur gears mounted on a common axle with their teeth out of phase, an anchor-shaped pallet rotates between them. The ends of the pallet alternatively catch and release the projecting teeth on either gear.

The first known European powered complex gearing appeared in Giovanni di' Dondi's mechanical clock of 1364. These medieval clocks were the first machines to be made entirely of metal. He had spent sixteen years constructing his clock in Padua, Italy. The result of his work was a machine technically far ahead of its time. The planetary motions provided the feasts of the church calculated with a perpetual calendar and the phases of the moon. The astronomical motions required seven dials with gears and linkages. Unfortunately, only one replica has made it into the present time. The gear train included elliptical gears and a sun and planet planetary. The brass and copper gears were cut by hand with the aid of some dividing instrument that ensured pitch accuracy.

Figure 3-7 Henri De Vick Clock

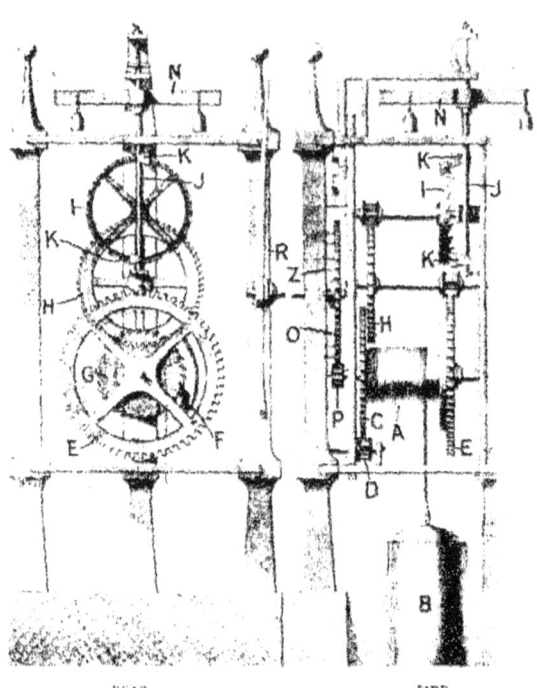

A 1370 clock (figure 3-7) built for the Palais de Justice by Henri de Vick took six years to complete. The parts included gears as large as three feet that had been made by a blacksmith using an anvil and hammer. The drawings were published in a widely known treatise by Moinet. The primary parts of the clock were also identified in a poem by Foissart.

Figures 3-8 and 3-9 illustrate typical English clocks of the period. The former shows a clock that was made by Peter Lightfoot and it would remain in use for over three hundred years. The clock operated from 1302 until 1835 when it was preserved and is presently exhibited in the South Kensington Science Museum.

Fig. 3-8 Glastonbury Clock Mechanism

Fig. 3-9 Clock Mechanism Wells Cathedral c. 1390

The differential gear reappeared in Eberhart Baldewin's 1575 astronomical clock built for the court of Wilhelm IV, Kassel, Germany. This clock is now in the British Museum. Isaac Newton developed a man powered filing machine for cutting clock wheels, and a formed wheel for cutting spur gears. Machine tools for the production of screw threads and precision parts developed simultaneously with clock making.

In Paris, Antoine Thiout published "Traité del'horlogerie" in1747, providing the first evidence of a clock maker with a knowledge of gear tooth mathematics.

Machine Tools: The lathe is of an unknown age and it has been credited with being the oldest machine-tool. A "Medieval Handbook" from Southern Germany, author unknown, circa 1480, has amongst the sketches a thread-cutting lathe with a single slide rest, a tool-holder and a feed in a dove tailed guide and is complete with a replaceable cutting die. During the 16th century the art of lathe making made further advances. Screws for fine adjustment that ensured the making of more accurate machine parts could now be produced. The lathe was the basic machine tool from which all future machine tools were derived. Even so the lathes were still crude, made of wood, and mostly 'pole" lathes, operated by a cord extending from a foot treadle, to the work, and then to a spring board secured to the ceiling. (Fig. 3-10)

Europe would still be using the pole lathe, the earliest known lathe form, in the 17th and 18th centuries. The work turned between centers while the operator held the tool against the work. Throughout the 17th century lathes would be adapted by the use of cams and patterns to produce irregular shapes. French metal cutting lathes were used in the mid 16th century. At the beginning of the 18th century we know of a metal-cutting lathe that was made in 1701. The first heavy industrial lathes would be built in France by the French machine builder Jacques de Vaucanson between 1770 and1780. Machine tools had shown negligible improvement in the period from 1400 to 1700. A print in Latin with a description is in the possession of the Astor Library, N.Y. The print details a screw-cutting lathe as shown in figure 3-10, and is reproduced from a previous work by Besson that was first published in 1569.

Fig. 3-10 Foot –powered Lathe in 1395

Jacques Besson was a French mathematician, inventor and engineer and was probably the first to publish his mechanical inventions. Besson produced a number of important treatises on machines, *"Théâtre des Instruments et Machines" in 1569* and *"Méchaniques"* in 1578, and *"Machinery"* in 1582. The latter illustrates a crude right angle bevel and worm gears. These were among the first books to be machine printed. As previously noted these early books had a similar style in that they included a full page sketch and limited text. Besson's contribution to the lathe in 1550 was to make it able to screw-cut metal. The lead screw controlled a tool firmly fixed to a support, the early slide rest. The machine could cut screws of any pitch or either hand. He also constructed slides for turning irregular shapes.

Fig. 3-11 French Screw-Cutting Lathe

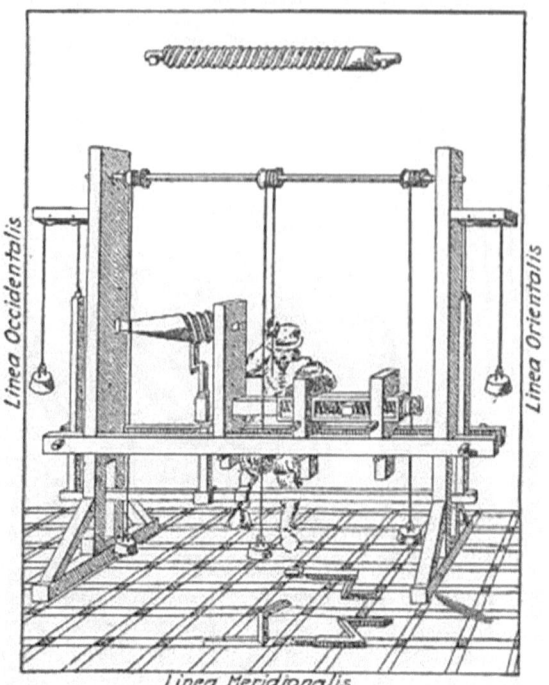

The circa 1600 manufacturing shop shown in figure 3-12 illustrates a lathe that is virtually unchanged over the previous two hundred years when compared to those in figure 3-10 and 11. These primitive lathes were used to grind lens contours by matching to a contoured gage. The free hand method was replaced by using turning operations in which the tools turned and ground using a metal tool termed a "*lap*". This tool was also produced on a lathe with a pivotal boring bar that could cut a reasonably accurate portion of a hemisphere.

Fig 3-12 Machine Shop Circa 1600

In 1675 Louis X1V's directed The Académie des sciences to undertake a study of machine principles and their application. A treatise was to be created covering all machines used by craftsmen and manufacturers. Jacques Buot the appointed coordinator died soon after the start of the project. It was to languish in Committees until a competing work, Dennis Diderot's "Encyclopédie" first volume was printed in 1751.

Gear Cutting: The first use of a gear cutting machine took place in 1540 at Charles V observatory in Spain. It was developed by Juanelo Torriano for a clock that contained 1800 gears. Torriano came from Cremona in Lombardy. The teeth were made with a rotary file cutter on a machine of his invention. Three gears were cut every day for three and a half years, "no wheel was made twice because they came out right every time."

The oldest existing gear cutting machine is in the London Science Museum dated 1672. Probably Hooke's machine as it was mentioned in his diary. It is made entirely of metal. The formed cutter is driven by a hand crank through a gear train. A screw adjusts the cutting depth. The blank is indexed by use of a dividing plate. In 1697 Nicholas Bion was building "gear-cutting engines" in Paris, France, utilizing formed rotary cutters.

In Stiernsund, Sweden, Christopher Polhelm, or Polhammer, scientist and engineer established a mechanical laboratory to study machines in 1697. He operated a manufacturing plant with approximately one hundred employees to make a variety of metal products using water power. In 1699 Polhelm built a machine that could mill metal plates and produce profiles. Several different machine tools were built including those for cutting teeth. In 1708 he designed machine tools for producing cog wheels. He also produced a water powered iron cutting lathe in 1716 with a screw drive to the cutting tool. Polhelm made a hand-operated production machine to cut gear teeth with reciprocating broaches in 1729. The earliest use of broaching of which we are aware.

Figure 3-13 Polhelm's Gear Cutting Machine

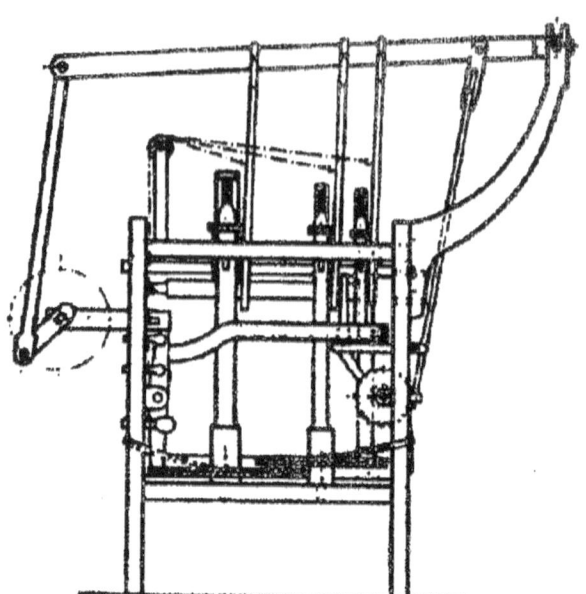

Soon afterwards this Swedish engineer introduced a power-operated machine to cut the teeth with a rotary cutters. Polhelm was also the first to build a fully automatic gear-cutting machine. I.e. from the first contact of the tool with the blank until all teeth are cut. The machined parts were of such precision that they were interchangeable. It would be 1860 before automatic gear-cutting machines reappear.

Gear Applications: Filippo Brunelleschi commenced building the famous Florence dome that bears his name in 1420. He is also famous for his machine inventions. His innovations included overdrive transmissions for winches and change gears to alter direction and speed. For the great crane (colla grande), Brunnelleschi, to reduce friction and prolong gear life, built pivoted gear teeth fixed in the wheels with metal pegs. Sources refer to these oak or bronze teeth as palei. Driven by a pair of oxen two horizontal gear wheels drove a vertical wheel alternatively, a worm gear arrangement raised or lowered the drive gear. By means of gears the horizontal shaft could handle different loads at three different speeds. The crane had a safety device to prevent the shafts running in reverse. Da Vinci analyzed his work and the use of low-friction gears with rotating cylindrical bearings.

Francesco di Giorgio Martini, was one of the outstanding Siena artist-engineers. He built many gear operated machines that included worm and rack gears. (1439-1501). Francesco was influenced by Taccola and included a number of Taccola's sketches in his earliest notebook. He also and provided detailed descriptions of eighteen machines and describes fifty-eight varieties of mills and the type of gearing that is required for each. Part of his work in "Trattato 1 and 11" was translated and published in 1967.

In his description of gears he specifies the shape and the best number of rods to be used in his lantern wheels. "On the shaft of the wheel is a lantern gear (two feet in diameter with 16 round rods), which rolls over the vertical teeth placed around the top of the horizontal crown wheel. In the back, another lantern gear is attached to the shaft, which turns the mill stone. This lantern gear rolls on the horizontal teeth of the same wheel." In this unusual arrangement the main gear wheel had two sets of teeth at right angles to one another. These gear wheels were driven by a horizontal lantern pinion and at 270° drove a vertical lantern wheel pinion.

Fig. 3-14 Francesco di Giorgio's Dredge Design

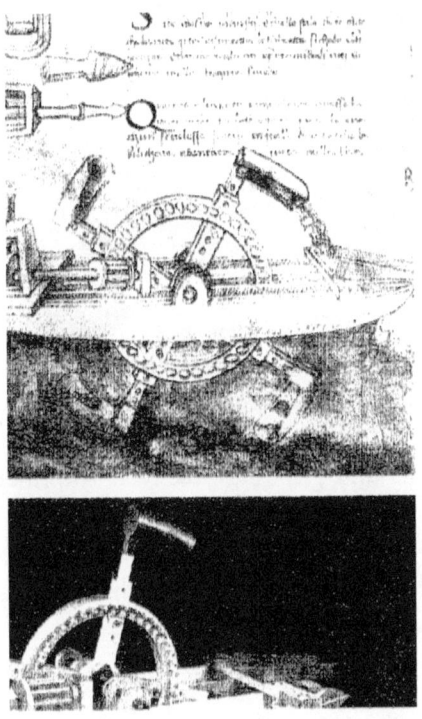

di Giogio wrote that gear strength was influenced by the tooth size and spacing. He also gives an explanation of the reason for a wheel turning easily as due to the difference in the tooth size between the crown wheel and the size of the lantern gears, and also provided a clear description of the differential.

In 1472 Roberto Valturio's treatise "De re militari libri X11" was published and is believed to be the first engineering book to have been printed. The book became the most popular military book of the 15th century. Figure 3-15 illustrates a war vehicle as described in the book, powered by a wind sail and driven through a gear train on both sides. A similar concept to Vigevanno' first armored vehicle of 1332.

Fig. 3 -15 Valturrio's War Vehicle

The Württemberg engineer and architect Heinrich Schickhardt (1558-1635) was another mill designer with an interest in gearing. His design of a complete fulling and grinding mill near Montbélard provides even more information than that provided by Sangallo and is an early example of perspective drawings. In each part of the drawing gear ratios are noted but critical gear information is omitted relying on the builder to be able to work within the envelope supplied. To provide the Hellenstein castle with water that had to be raised ninety meters he designed and had a carpenter build a combination of a lantern and oval gear rack for the pumps.

In a 1659 Paris exhibition machinery fitted with cycloidal toothed gears could be seen.. Two years later we know the French mathematician Girard Desargues built machinery utilizing the first epicycloidal toothed gears. Only with his work did it become certain that the geometry of transformations was based on valid principles of projection. In the development of mathematics the new mathematical discipline, projective geometry, is credited to Desargues as its founder.

We have previously written about Jacob Leupold.s (p.52 and later p.88) treatise on machinery in 1724. Largely devoted to the engineering of wind and water mills he dealt in detail with lubrication, bearings and gears. In Leipzig, in 1695, Leupold observed the building of an ingenious gear driven jack. The two inch pitch rack was traversed by a six pin gear wheel. The pins were equally spaced on a twelve tooth worm wheel driven.

During their occupation of Bolivia the Spanish built mints for the processing and marking of silver with the King of Spain's seal. It is estimated over 32,000 tons were shipped to Spain between 1531 and 1660. (World production in 2004 was 16,117 metric tons.) The Spanish piece of eight became the first world currency. The Spaniards used a crude gear technology in their South American mints as can still be seen in existing machinery (Fig: 3-16).

Figure 3-16 17th Century Cas de Moneda Mint, Potosi, Bolivia

Gears for Mining: In Saxony, George Bauer's, Latin name Georgius Agricola, monumental work "De re Metallica" was posthumously published in 1556, an illustrated twelve volume encyclopedia on mining and metallurgy. It was first published in Latin and translated by President Herbert Hoover in 1912. A practicing physician, Bauer was the first systematic mineralogist. The books provide many details on the mining machinery of the period and contain etched double-page illustrations. Two hundred engineered projects such as pumps, cranes, mills, bridges and ballistic equipment are described in great detail. For 180 years his book was the standard metals text book and guide until superseded by Schlütters work in 1738. This early printed book on machines illustrated a wide variety of gear applications. Towards the end of his book machines that appear to be impracticable are illustrated. bars in a lantern pinion.

In figure 3-17 Bauer illustrates a rundle drum drive. The two to one speed reduction is achieved with pinion C constructed as a rundle drum. The drum is composed of rundles, a rundle being any of the bars in a lantern pinion. The pinion drives a sixty tooth wheel, the teeth are a foot in length. The teeth Hare triangular shaped and glued in place. With a short arm the ratio of the load to crank force was calculated at 62:1 which doubled with the long arm.

In figure 3-18 the drive was described as: "... a smaller toothed wheel, which has forty-eight teeth... Around the third axle... a seventy-two tooth teeth. The teeth of each wheel are fixed in with screws, whose threads are screwed into threads in the wheel, so that those teeth that are broken can be replaced by others. Both the teeth and rundles are steel." This machine is a prototype of our present day screw conveyor. This is also the first time that we know of when a gear box made entirely of metal is described. The iron axles revolved in metal bearings. The teeth and rundles were also made of steel. Individual teeth were screwed into the wheel for ease of replacement.

Fig 3-17 Rundle Drive

Fig 3-18 Underground Mine Draining Machinery

The Italian military engineer, Agostini Ramelli, described himself as engineer to the most Christian King of France and Poland. He continued Besson's work and in 1588 published his only book "Le Diverse et Artificiose Machine", in Paris. Unlike Besson's bevel gear, Ramelli's gears were more highly developed and suitable for operating at any angle (ref. fig. 3-19). In this machine he makes use of three Archimedean screws. Gears were cast solid or made of wood with removable metal teeth. Lantern pinions had round bars for driving. The teeth faces were not on an epicycloidal curve and were of low efficiency. Ramelli realized that a small pinion with a large wheel increased the torque.

Fig 3-19 Raising Water by a water Powered Ramelli Machine

The sketch figure 3-20 is of hand operated worm gear expanders and can be considered illustrative of the crude gears of Ramelli's period. It is also noteworthy in that the racks are supported by rollers. Some of the pumps shown in this important book were lowered and raised vertically by means of racks. Ramelli's brilliance is illustrated by his compounding two water wheels using a vertical overshot wheel. The water was retained in a reservoir and released on to the blades of a horizontal wheel. The power was directed to the same system, and released to two different machines.

Figure 3-20 Hand Operated Worm Gear Expanders Ramelli Period.

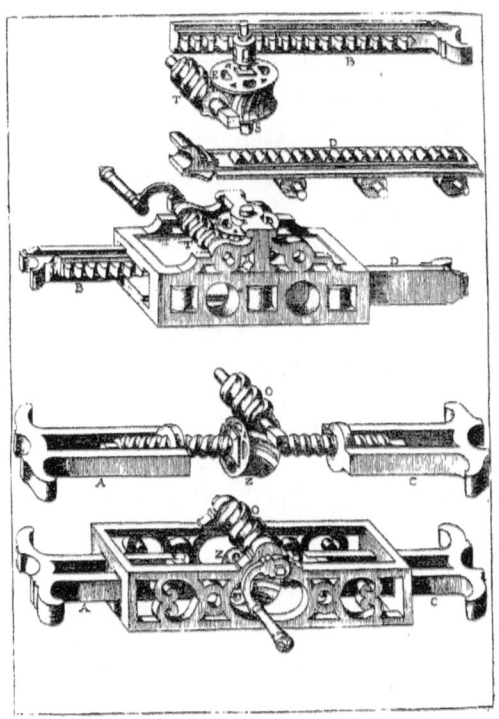

Some thirty-seven years earlier the learned judge and director of the Istanbul Observatory, Taqi al-Din, had in Damascus written in Arabic on similar devices to Ramelli's. By this time Latin had largely replaced Arabic as the universal language of science. Al-Din has been credited with at least nineteen books on light and marvelous machines. A catalog dated 1531 described clocks, pumps, and lifting devices he had designed. In many 16th century applications gear wheels were constructed like wooden wagon wheels, or boards laid alternatively transverse to those above and below. These wheels carried hardwood teeth mortised into the rim or face making them easy to replace when worn.

Another important contributor to descriptions of 17th century industrial machinery was Vittorio Zonca, a Padua architect. In his 1607 book "Teatro Nuovo di Machine et Edifici." Zonca revealed enthusiasm for the development of mechanical machines, referred to wear and the consequences of running steel on steel advocating steel on brass. Shown are heavy metal gates opened with the aid of a geared driving shaft. The solid metal gear wheels with a central square hole indicate advance construction.

Grinding: Tool sharpening mills were built in England as early as 1405. Two large mills were built in 1428 that were forbidden to convert the mills to any other purpose than sharpening newly manufactured edge tools. Enormous quantities would have been sharpened to defray the cost of the mills John Payne, of Glastonbury, England, obtained a patent in 1573 for a grinding mill. The spindle carries a "trendell" in gear with "the cogwheel" on the "spyndell".

At this time mill stones were chosen from several natural stones and were dressed by hand into one of several designs as shown in figure 3-22. Small items were hand ground on a machine similar to what is shown in figure 3-21 that has existed to the present day.

Fig. 3-21 Treadle Grinder.

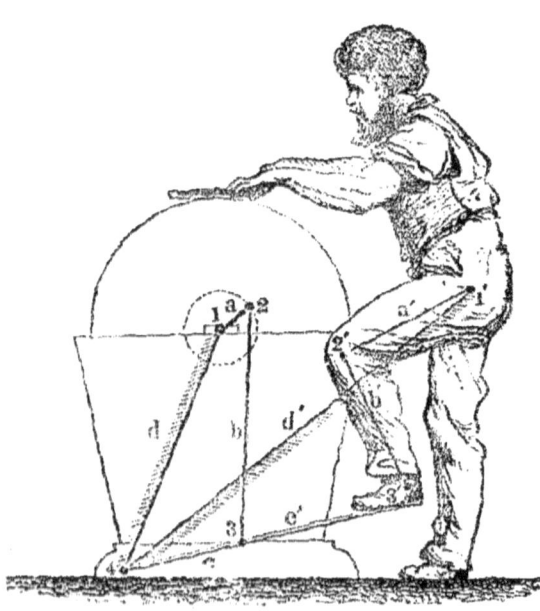

Johannes Stradanus's 1575 copper engraving from Berlin shows an armor grinding and polishing shop. These belt grinding machines made no further advances that can be determined until the end of the 18th century.

Grinding stones were used for three purposes, to smooth surfaces, reduce metal to a required thickness, and sharpen edge tools. The stone surface used for sharpening tools was required to be smooth, and the surface to remove metal as quickly as possible was termed "hacked".

Fig. 3-22 Early Grinding Stones Dressing Designs.

(A) Late Roman (B) 18th Cent, (C, D, & E)19th Cent.(F) Right-Hand (G) Left-Hand.

In Muhlen, an engraving was made by G. Strada in 1629 and included in "Kuntstliche Abriss allerhand". The engraving depicted a perpetual motion grinding machine as shown in figure 3-23 with its complicated gear arrangements.

Fig. 3-23 Worm Device for Grinding Wheels

Power Sources: From documentary evidence we know the early mills were powered by either animal, human, water or wind. The upper of a single pair of millstones had to revolve at a speed that could grind the material, usually grain, and expel the ground meal. The motion of the vertically mounted wheel was transferred through ninety degrees necessitating gearing. A wooden cog-wheel or trundle-wheel was mounted vertically on the axle's inner end meshing with a lantern-pinion on the vertical spindle that was inserted in the stone. By the 13th century the vertical mills were replaced by horizontal mills. Present day calculations indicate the power of a Roman donkey or slave at half-a-horsepower, the horizontal wheel at about the same, the undershot vertical wheel in the range of three horsepower, and the medieval overshot wheel from forty to sixty horsepower. One man with two horses could do the equivalent work of five men. Leupold provided an appendix on measuring power. The weakness in generating power lay in the use of wooden parts that were easily replaced. Due to the high cost of iron metal gears were not considered until after the eighteenth century.

The first windmill in N. America was installed at the Flowerdew Hundred Plantation on the James River, Virginia, in 1621 to be quickly followed by many thousands of similar installations. Windmills were driven in the reverse of waterwheel mills. The windmill situated the horizontal shaft at the top of the structure geared at right angles to the millstones below. Waterpower frequently required the construction of dams. The increased use of water power resulted in many law suits after 1300 on the rights of water-power over navigation.

The Italian, Giovanni Branca, in 1631 built the first practical steam engine advancing the work of Giovanni Baptista della Porta (steam pumps) and Valturrio's pupil Solomon de Caous work of 1615. The steam driven wheel operated a gear train to alternately raise two weighted arms. In 1698 Thomas Savery was issued a patent for a "new invention for Raiseing of Water and occasioning Motion to all Sorts of Mill Work by the Impellent Force of Fire." Savery, English inventor and military engineer, demonstrated the first useful steam-engine in 1698 to William 111 at Hampton Court. The patent was so broad it restricted almost any new invention associated with the steam engine. In 1712 the Newcomen engine was sold under Savery's patent.

In 1650 the German engineer Otto von Guericke had shown atmospheric pressure could be used as a force with his famous demonstration of the Magdeburg hemispheres. Guericke was also the first to build a machine for producing an electric charge. Building on his work on air pressure The French physicist Dr. Denis Papin at Marburg discovered the true solution – the condensation of steam in a cylinder fitted with a piston. Papin made a working model of an atmospheric condensing steam engine in 1690 whose principles were later developed by Newcomen and Watt. This is considered as the birth of the steam engine. Papin also assisted the Dutch physicist Christian Huygens and later Robert Boyle. Huygens described the theory of an internal combustion engine in 1678.

The transformation of Newcomen's engine into an efficient and practical power source suitable for factories and eventually ships and locomotives was due to Watt. In 1764 he converted the previous jolting motion into a smooth rotary action. Other improvements were the elimination of the alternate heating and cooling. An additional cylinder allowed for the condensing of the steam without affecting the engine performance. The engines were low pressure and steam engines only became compact and efficient wityh the development in the USA of high pressure steam engines by Oliver Evans.

Ultimately the steam engine would be the driving force behind the industrial revolution in both transport and factories, liberating slaves, peasants and animals from drudgery. The next major advance in gearing would also be created by the demands of the steam engine, to be followed by the needs of wind power, the bicycle, automobile, helicopter, and advances in power generation. The latter currently requiring that gears run with tip speeds of 200 m/s and powers to 140 MW. In the seventeenth century the most powerful prime mover was the hundred horsepower drive for the Versailles waterworks pump. The pump was used daily to raise a million gallons 502 feet.

Advances in Measurement and Calculations: In 1617 John Napier the Scottish mathematician and inventor of logarithms, described the mechanical device for multiplication that became known as Napier's Bones or Rods in his "Rhabdologia". By moving several rods, each with an appropriate set of numbers to show a pair of factors the product could be read from other rod numbers. His invention of logarithms was described in "Mirifici Logarithmorum Canonis Descriptio" (1614).

The French scientific instrument-maker Pierre Vernier invented the popular auxiliary scale that we know as the vernier to provide accurate readings in 1630.

William Gascoigne c.1612-1644) of Yorkshire, England, made the first micrometer in 1632. Two parallel edges were moved by means of a screw with a dividing head. The micrometer measured the angular distance between stars on a telescope. Huygens made further improvements in 1657. It would be more than two hundred years before Brown and Sharpe impacted the market with accurate and readily available micrometers that were based on the Palmer micrometer they had brought back from France.

The first mechanical calculator that could add and subtract and also the first to use gears was patented by Blaise Pascal, at the age of nineteen, in 1642, Pascal calculators were similar to those in use in the 19th and early 20th century. They utilized gear ratios; however, machinery was not yet available to make accurate gears. Over a ten year period, starting in 1642 Pascal built fifty different versions of his calculator that were called pascalines. The calculators contained between five and eight interlocking gears in a brass box. They were exhibited and offered for sale in Paris in 1652.

An alternative method for varying the number of teeth in engagement was known as the "pin-wheel". Described in 1709 It was known to Leibniz and in use for calculators in the 19th century. The wheel had nine movable teeth that could be disengaged or any number engaged with an external wheel in order for the counter to advance forward.

In 1664, Henry Power had published in London "Experimental Philosophy" with the first account of microscopic examination of metals. C.S. Smith in his "History of Technology" provides a quotation by Power's "Look at a polished piece and you shall see them all full of fissures, cavities, and asperites and irregularities." Robert Hooke published his findings in 1678 under the title "De Potentia Restitution" which included the basis of what we know today as "Hooke's Law; "In a metallic substance the amount of strain resulting from elastic deformation is proportional to the stress applied to produce it" Hooke also developed the microscope and made observations on metals.

Two books were published by Sir John Pettus in 1670 and 1683 respectively that provided the general information available on both ferrous and non-ferrous materials. At this time he was Deputy Governor of the Mines Royal. His "Glossary of Metallic Words" provided much of this useful information.

John Moxon, a Fellow of the Royal Society, published the first textbooks on technical subjects. In his 1678 book "Mechanik Exercises" he gave a poor opinion of English and Spanish iron that was likely to crack between hot and cold. He provided the sources for imported metals and used advanced terms such as "fine grain". His detailed case-hardening recipe for iron included a mixture of powdered dried horns and hooves, salt and stale wine that was virtually unchanged since Theophilus.

In 1687 Sir Isaac Newton's "Principia" was published presented his hypothesis on viscous flow which became the basis for the theory of fluid-film lubrication.

The Bernoulli brothers and Huygens studied the function of the brachistochrone and tautochrone curves.. Huygens discovered the solutions in the cycloid and in 1673 used the information to construct tautochronous pendulum clocks. In these clocks the period is independent of the amplitude.

Metric Measurement: In 1670, Abbé Mouton of Lyons, France, made a proposal for a basic metric system. At the instigation of Talleyrand the French Academy formed a committee in 1784 to devise a new system of measurement t/hat would become the ISO metric system. The third American president, Thomas Jefferson, proposed a highly developed decimalization system. In 1866 the U.S. passed an act establishing metric measurements (Act 14: Stat. 339; U.S.C. 204). The U.S. was an original signatory of the metric 1875 Treaty of the Meter (20 Stat 709; USC 205). This treaty established the tables that may be lawfully used to compute, determine, and express customary weights and measures in the metric system.

The U.S. Metric Conversion Act of 1975 (P.L. 94-168: 15 U.S.C. 205) established that ":It is the therefore declared that the policy of the United States shall be to coordinate and plan the increasing use of the metric system and to establish a United States Metric board to coordinate the voluntary conversion of the metric system (15 U.S.C. 205b). The act was amended on August 23rd, 1988 (P.L. 100-418) and now states in part "It is therefore the declared policy of the United States:

1. *To designate the metric system of measurement as the preferred system of weights and measures for United States trade and commerce.*
2. *To require that each Federal Agency by a date certain and to the extent economically feasible by the end of the fiscal year 1992, use the metric system of measurement.*

Chapter 4

Progress Through the 18th Century

General -- Metal – Mathematics/ Physics – Thermometers – Power – Technology - Involute – Applications -- Machine Tools - Tooth Cutting

General Developments: In the 1700's the lay-out of a manufacturing shop with its tools revealed minimal advancement over what could be seen in the mid 1500's. Many historians believe that the industrial revolution started in England in the mid-part of the 18th century with extensive agricultural innovations and textile factories. Industry gathered momentum with increased mechanization and the use of steam engines.

The beginning of truly scientific engineering was largely revealed in the writings of Jacob Leupold and Bernard Forest de Bèlidor. The latter (1697-1761) was understood to be the leading engineering writer of the period. His algebra based books on mechanics linked the earlier works of Galileo and Stevin to practical engineering projects. With the use of steam engines by the early 19th century the change from an agricultural to an industrial economy would be complete.

Prior to the 18th century European engineering was the work of craftsmen, monks, military and amateurs, whereas the achievements from now on would be by engineers and scientists. The century would see a rapid advance in the understanding of the relationship between mathematics and its application to mechanics. Foremost in this work were the Bernoulli brothers, Euler, Leibniz, Lagrange and Laplace. Physics and technology also made considerable advances. The engineering profession was also one of the great developments. By 1771 there were enough engineers in England to be able to hold regular meetings at what they named "The Smeatonian Club".

Improvements in agricultural and textile manufacture were advanced with newly designed threshers and looms. The latest power source, the stationary steam engine, began to show its value. The introduction and gradual increased usage of the steam engine was the most important development of the 18th century. Also an imperceptible replacement of wood with iron was taking place.

Practical engineering text books became generally available. Born in Planitz, Saxony, Jacob Leupold referred to himself as a mechanicus – a professional with a solid theoretical grasp of mechanics. He applied this knowledge to supply an engineering text for practicing engineers. Leupold wrote between 1723-39 "Theatrum machinarum generale" in ten volumes, the first systematic writings on mechanical engineering. He was the first writer to separate single mechanisms from machines. The last three volumes of this engineering encyclopedia were published posthumously. The encyclopedia was written for practical application by those artisans who had not received a formal mathematics or science education. James Watt (1782) studied this work, and the great engineers John Smeaton and Thomas Telford acquired all ten volumes. The work included the design of a non-condensing, high-pressure steam engine that was comparable to those that would be built a hundred years later and included information on gears. In 1727 Leupold built a nine dial calculating machine using ten tooth gears. He also wrote on the subject of lubrication.

The first volume of "Encyclopédie, ou Dictionnaire Raisonné des Sciences, des Art et des Métiers "was published in France in 1751--1776. This great encyclopedia of Denis Diderot and Jean Le Rond d'Alembert contained eleven volumes of plates detailing the manufacturing methods of every known trade. The encyclopedia became the most influential publication of the "Enlightenment", the 18th century European philosophical movement, creating an impetus toward learning. Twenty competing volumes approved by the Académie des Sciences were finished in 1781.

As British industries multiplied so did the demand for iron. In 1750 iron would only be used when necessary with castings having less than five percent. By 1830 iron would be the material of choice for engineers. A lobby was formed to ensure that British industries did not lose their market to the colonies. By 1750 the Iron Act had been passed forbidding the erection of slitting mills, rolling mills, and steel furnaces in the American colonies. The English Parliament, in the mistaken belief that they were protecting their industrial leadership, passed a law in 1785 banning the export of all iron and steel machinery, their plans also intended to stop the emigration of mechanics and workmen in the iron and steel trades. Severe penalties were included for anyone breaking the law. The result over the next several years was a large exodus of labor with machine designs to North America. These laws aggravated the colonists. American ingenuity in designs, and their skill in evading laws that hindered them, would result in the rapid growth of American industry.

From 1730 to 1790 the whole structure of industry had changed. Plentiful supplies of pig-iron and production quantities of tougher, malleable iron castings were available thanks to Darby and Réaumur. Easily adapted bar iron could be obtained due to Henry Cort, and improved steel for cutting tools was being used courtesy of Huntsman. Cast-iron cylinders could be accurately bored and steam engines using coal had become practical.

Metals: Abraham Darby, an English iron-master, took out a patent for his casting process in 1708. He developed sand-casting, and in 1709, at Coalbrookdale, Shropshire, England, was the first to successfully smelt iron with coke in a blast furnace, making the industrial revolution possible. His son, Abraham Darby 11, was successful in forging cast iron in 1748, by using an ore low in phosphorus. He discovered how to produce wrought iron from coke-smelted ore, but kept the process a close secret. A modern historian of this period's iron making wrote" it is true to say that the industrial revolution started here." By here he meant Darby's "Old Furnace".

Coal's sulphur content was too high; charcoal was too soft and scarce, and was crushed under the weight of the ore. This important breakthrough of using coke was not publicized and in the next fifty years only six other coke furnaces were built in Britain. In 1770 the use of coke was more common but still not universally used. His son and grandson of the same name became famous in their own right for achievements in the production of metals. The grandson, Darby 111 built the first cast iron bridge, that is still standing at Ironbridge, Shropshire. The most historic man made structure since the cathedrals of the middle ages. It is a beautiful structure with circular symmetry and a gossamer framework towering one hundred feet over the river Severn.

John Roebuck, a chemist and the Carron Iron Works founder converted cast iron into malleable iron in 1762.s for making wrought iron. After Roebuck's early death Henry Cort, The Great Finer, obtained the patents. English iron-master Cort in 1784 patented the wrought iron Puddling Method using coal in place of charcoal. The puddling process converted brittle cast iron into wrought iron. Cort burned the coal in a separate part of the furnace for more efficient production by excluding sulphur and carbon from the iron. Even so the process was inefficient and wasteful; half the pig iron was absorbed by the slag. The resulting blooms were hammered to get rid of the slag. Cort also patented a rolling process, patents 1,351 in 1783 and 1420 in 1784. His first patent was related to the hammering, rolling and welding of iron, the second patent was for a reverberatory furnace with a concave base. Iron could now be made consistently by the ton. In 1816 Joseph Hall of Tipton, Staffordshire, introduced an improved method superseding Cort's and would be put to use in all other industrial countries. Cort was to die poor and stripped of his patents due to his partner's embezzlement.

Metal workers knew only two hardnesses, either capable of being filed or not. Several kinds of stones or similar hard materials served as files. The Moh's system was thus predated by a hundred years. This and other test methods were all described in detail by the French physicist René-Antoine Ferchault de Réaumur. Member of the Académie Royale des Sciences, he was one of the few scientists interested in the problems of engineers. He oversaw an official description of the arts and crafts. Born in 1683 he was the first to make a scientific study of the nature of iron and steel. He discovered the role of carbon in steel hardness, and also devised the first method for evaluating hardness in 1769. His 1722 book *"L'arte de converter le fer forge en acier"* was the first technical treatise on iron and steel. The second part of the book dealt with the process associated with his name that produced soft malleable castings of white iron. The castings were annealed in a mixture of charcoal, bone ash and chalk. Réaumur was the first to investigate fractured surfaces. Using only a hand held lens he drew what was revealed, illustrating the features in his *"Memoires"*, methodically using the fracture appearance as a quality test. This latter book was translated by A.G. Sisco and published by the Chicago Press in 1956 as *"Memoirs on Steel and Iron."* He edited another important book dealing exclusively with iron and steel *"Description des Arts et Métiers".*

Polhelm was the first to use large scale rolling mills for working metals and understood quantity production economies. (See p.60). In 1734 the prolific Swedish writer, Emanuel Swedenborg, theologian, scientist, collaborator and former pupil of Polhelm, wrote on technology and engineering. Mining. is described In his "Regnum subterraneum sive minerale de ferro Philosophica et Mineralia" written in 1734. He wrote in his book "De Ferro" an account of circular cutters being used for slitting mills in the Liége district, Germany, England, and Sweden.

Figure 4-1 Clipping Mill

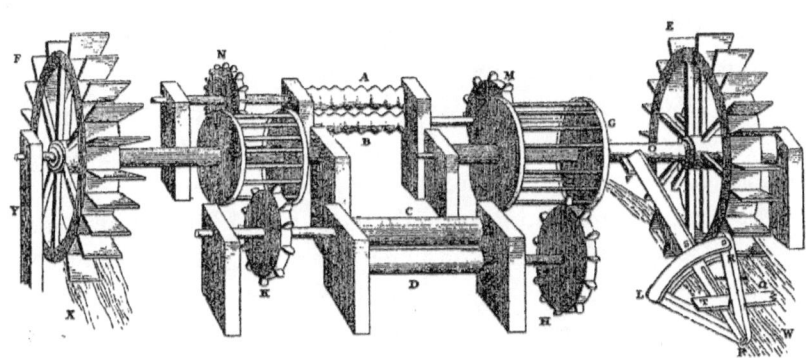

In Emerson's "Mechanics" published in 1758 is the earliest known drawing of an English iron work mill. As shown in figure 4.1. It included a slitting plate and clipping mill.

In 1742 the English inventor and clockmaker, Benjamin Huntsman, developed the crucible process in a Sheffield foundry. The method was sorely needed because of the variable quality in English steel. His method could only produce small quantities and his discovery was made empirically. It was of major importance in that the feeds and speeds of machine tools could take advantage of the improved cutting capacity of the carbon steel. Steel quality and characteristics could now be controlled. He patented his method for manufacturing cast-steel. At the time cast-steel was the most uniform in quality, the hardest steel, and the best material for cutting tools.

Some thousands of years ago it was known that iron with the right amount of carbon is capable of being hardened, and if in contact with charcoal would develop a hardened skin, a process that we know as case hardening. In the early years of the 18th century it was learned that the longer iron was heated when packed with carbon the deeper the carbon penetrated giving a thicker case. It became known as the cementation process. Iron was heated for as long as three weeks to obtain shear-steel suitable for textile shears. Huntsman re-melted shear-steel and found he could obtain the best steel available. The Swedish metallurgist, Torbern Olaf Bergmann, was in 1750 the first to realize the importance of the carbon content on steel hardness. In 1786, Joseph Priestley, one of the discoverer of oxygen, reached the same conclusion believing that iron and carbon chemically bonded. The process would not be improved upon until David Mushet's patent of 1800. Of all the papers published to date on the manufacture of iron and steel Mushet's "Papers on Iron and Steel" that had appeared in the "Philosophical Magazine" prior to their 1798 publication were considered to be of the most importance.

In his 1766 "Dictionnaire de Chimie" the French chemist Pierre Joseph Macquer was the first to describe the sintering process, and the scientific principles upon which powder metallurgy depends.

Mathematics and Physics: In 1712 the Ducal Library in Hanover published Gottfried Wilhelm von Leibnitz's paper that was important in differentiating between rolling and sliding friction. The French military engineer, Charles Augustin de Coulomb (1736-1806), for whom the coulomb is named, the unit of quantity in measuring units of electricity, being the first to apply mathematics to electrical and magnetic interactions. His work is also connected to the strength of materials, and the most comprehensive study of friction which concluded:

* Force of friction depends on force normal to the contacting surfaces.
* Force of friction is independent of the area of contact.
* Potential static friction is greater than kinetic friction.
* Kinetic friction is independent of the relative velocity of two bodies.

William Jones, who translated some of Newton's works from the Latin, introduced the symbol ϖ in his book "a New Introduction to the

Mathematics". It is suggested that he used the letter as an abbreviation for the word periphery. The symbol did not become established until used by Euler in 1737 in his "Variae observationes circa series infinitas", previously he had used p or c.

As the mathematics professor in the artillery school at La Fère, France, Bernard Forest de Bélidor published the first formal algebra-based mechanics text books. He was probably the most influential engineering author of the period. In 1725 his "Nouveau cours de mathematiques" was published, followed by "La science d'ingénieurs" in 1729.

John Winthrop established the first laboratory of experimental physics at Harvard in 1746. He provided demonstrations on the laws of mechanics, light, heat and the movement of celestial bodies. This brilliant physicist was elected Harvard professor of mathematics and natural philosophy at the age of twenty-four. In 1751 he introduced the study of calculus into the curriculum.

The French mathematician, Joseph Louis de, Comte Lagrange, was born in Turin. He succeeded Euler as director of mathematics at the Berlin Academy. Frederick the Great considered him the greatest mathematician of Europe. He became a senator under Napoleon and taught at the École Polytechnique. One of his most important works "Traité de mécanique analytique" was published in 1788 in Paris. Lagrange's methods were studied, tested and applied.

Adrien Marie Legendre published his text book "Eléments de géométrie", in 1794 which became the basis for modern geometry texts in American schools. The book was translated into English by Thomas Carlyle.

École Polytechnique the first polytechnic school was founded in Paris in 1794. Gaspard Monge, Professor of mathematics at Mézières, worked out the rules for what he called Descriptive Geometry in 1765. Previously, dimensional solutions were by laborious arithmetical methods. Monge was one of the founders of the new school and became the school's professor of mathematics. He was influenced by the work of Dürer and wrote "Leçons de géométrie descriptive" in 1795 describing the mathematical principals underlying complex graphical techniques. After his death the school continued to flourish in mechanics and mathematics. In addition to Monge and Lagrange, Ampère, Poisson, Malus, Fourier and Cauchy were some of their outstanding scientists.

Count Rumford, born Benjamin Thompson in Woburn, Massachusetts, was commissioned in the New Hampshire Regiment, entered the service of Bavaria, reformed the army, and was knighted for services to the British Crown. In 1799 he founded the Royal Institution, endowing Rumford medals at the Royal Society and the American Academy for researches into light and heat. He published work from which can be derived a quantitative measure of the mechanical equivalent of heat.

Thermometers: In 1654 the Grand Duke of Tuscany invented a thermometer that used liquid in a glass tube that had one end sealed. It was an improvement on Galileo's thermoscope, and would be improved by Fahrenheit sixty years later and become our modern thermometer De Réaumur made a thermometer using spirit dividing the freezing and boiling points by eighty degrees in 1708.

Born in Danzig in 1714, Daniel Gabriel Fahrenheit produced a mercury-in-glass thermometer which was vastly superior to any previous temperature gage. He devised the temperature scale named for him, and was the first to reveal that the boiling point of liquids changes under different atmospheric pressures. Fahrenheit had previously consulted with Réaumur on the subject of temperature measurement.

Anders Celsius of Sweden built his first thermometer with what we now know as the Celsius scale, the upside-down scale. The zero, the boiling point of water, was on top, the 100° freezing point, below. Jean Pierre Christin of Lyons produced the right-side up thermometers in 1743.

Power Sources: This would be the first century when power would be attainable other than by wind, water or muscle power, thanks to the work of Thomas Newcomen in collaboration with Thomas Savery.

Figure 4-2 Horse Powered Speed Increaser

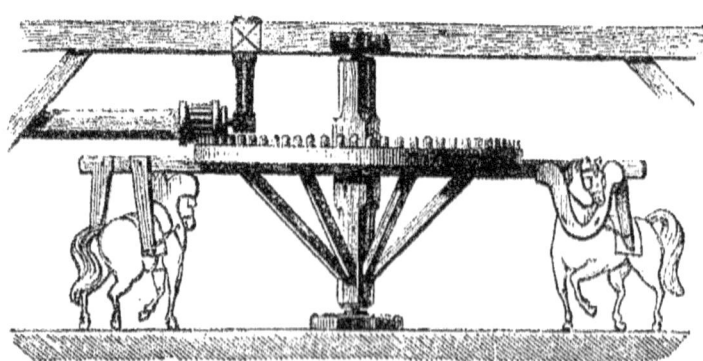

Savery, was an early pioneer in applying heat energy to drive pumps, in 1698 he had patented the first practical high-pressure steam engine to pump water from mines, and suggested the rate at which a horse does work should be the standard for the measurement of power. Savery estimated the average work horse pulled with 180 lbs of force on a capstan lever twelve feet in length producing a torque of 2,160 ft. lbs. The full circle was made within twenty-five seconds (0.251 33 rad/s). This rate times the torque equals 543 ft/lbs/sec. (32,580 ft. lbs/min). The term horsepower was first used by Scottish engineer James Watt and, the unit of power the watt, is named for him. He rounded Savery's calculation to 33,000 ft/lbs.

Engines: The first patent for a gas-engine was issued in Britain in 1794, thirty years later gas-engines would be in use for powering pumps. The steam engine provided the basics for the development of the gas engine. The first steam engine built for work and not the laboratory was Newcomen's low pressure engine. The engine was installed at Dudley Castle, Staffordshire, in 1712 to pump water from a mine and, was about 5.5 hp. This was a great improvement over Savery's engine and introduced a mechanical power source where previously human, animal, water and wind power had been used. This was the first practical steam engine to use a piston and cylinder. Savery's patent was taken over by a partnership that controlled steam engine manufacture for eighteen years after his death in 1715. By Newcomen's death in 1729 his engines would be in general demand throughout Europe until James Watt made them obsolete in 1775. The first Newcomen engine in the U.S. was imported by Josiah Hornblower for the Schuyler Copper mine, Belleville, N. Jersey in 1753. It was the first atmospheric steam engine to be used in the U.S. Prior to the Declaration of Independence there was only one other steam engine in the thirteen colonies and that was in a Philadelphia distillery, Britain having restricted the colonies from participation in most industries. The first steam engine in Europe would be erected at Kōnisgsberg, Russia, in 1722. A steam engine is also known to have been operating at a coal mine at Jemeppe-sur-Meuse, near Liège, Belgium in 1725.

Watt received an English patent for his first rotary engine in 1769. He transformed the inefficient atmospheric engine into a high-pressure steam engine by condensing the steam into a separate vessel. In 1775 he went into partnership with Matthew Boulton and was granted two more patents that doubled the engine power. Watt's engine converted the reciprocating motion into a rotary movement by using a planetary gear system. Watt wanted to avoid controversy in order to protect his own broad patents and for the first time incorporated a sun and planet gearing (epicyclic) motion which was also patented with his 1782 engine. In England, for the first time, Boulton and Watt's steam engine. that was to "Drive the Industrial Age", was used in 1780 to drive a forge hammer. The engine, Watt's refinement of the Newcomen engine, included a system of rotary gears that translated vertical motion into circular motion. His son, also named James Watt in 1817 was the first to fit his father's steam engine to power the first English steamship to leave port, the Caledonia. In 1785 Watt invented a parallel motion that was necessary because planers to machine the crosshead and guides would not be developed for another thirty years. The Patent Museum in South Kensington, London, owns several wooden models of Watt's proposed arrangements to obtain rotary motion from the beam, including the sun and planet engine which powered the machines in his Soho factory.

Figure 4-3 Watt's Engine with Epicyclic Arrangement

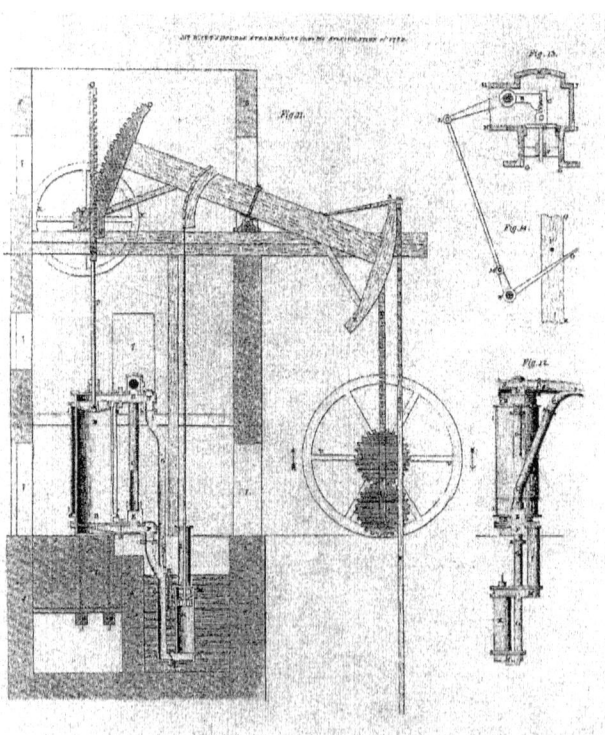

Watt used cast iron gear wheels with rectangular teeth. With wheels of equal diameter the planetary system provided the advantage of two revolutions for each engine cycle. Watt's refinement of the Newcomen engine, included a system of rotary and the disadvantages of being noisy and inefficient. Gears were rough un-machined castings. Another difficulty was that Watts did not have the capability of boring a six inch diameter by two foot long cylinder. The sun and planet gear changed piston motion into rotary motion. Producing rotary motion from reciprocating motion was a major achievement. The sun and planet motion had been independently invented by the Scottish engineer William Murdock in 1785.

Murdock also improved on Watt's engine, and pioneered distillation of coal gas. Watt is also credited with originating the micrometer and his instrument is also is in the Patent Museum, and he invented a steam engine pressure gauge. In 1798 Murdock took charge of the Soho foundry. As we report later he made many innovations to their machine shop. By the turn of the century numerous builders entered the steam engine market to take advantage of the expiration of Watt's patents. Richard Trevithick and Oliver Evans in the U.S. were predominant and built their engines without a condenser, exhausting the waste steam into the atmosphere. Trevithick was the first to build a practical high-pressure steam engine.

Experiments with steam engine vehicles began in France when Nicholas Joseph Cugnot was the first to apply a high pressure, or non-condensing engine. Except for one experimental vehicle which Cugnot had built, the vehicle in figure 4-4 was the first to be self-propelled.

Figure 4-4 Cugnot's Steam Powered Vehicle

This four passenger 1765 steam powered vehicle could travel at two mph. It was also built to carry artillery. This it never did, but it ran as designed and is now in a Paris museum. This is perhaps the most **important vehicle in the whole history of the motor car.** The single front wheel was powered through a toothed wheel on the front axle. The boiler could be swung with the guiding and driving wheel by a rack and pinion on a perpendicular shaft to the engine. Following this vehicle Cugnot built a vehicle with a crude steam engine powered artillery tractor.

Water & Wind Mills: As early as the first century wind was used by Hero to power a musical organ. The first practical windmill, used to grind corn, was during the reign of the Caliph Umar in Medina (A.D. 584-644). The windmill had large rectangular blades. In the 12th century they could be seen all over northern Europe. The windmills could usually deliver more power than the water mills and could be set up anywhere there was a suitable wind.

John Smeaton was a mathematical instrument maker. In 1753 Smeaton won the Copley Medal for his researches into the mechanics of waterwheels and windmills. His research was presented to the Royal Society in 1759 under the title "Experimental Inquiry Concerning the Natural Power of Water and Wind to Turn Mills and Other Machines Depending on Circular Motion." On April 25^{th,} 1776, Smeaton read before the Royal Society his paper titled, "An Experimental Examination of the Quantity and proportion of Mechanical Power necessary to be employed in giving different degrees of velocity to Heavy Bodies from a State of Rest." The efficiency of the water and wind powered mills was studied and improved.

The great era of the windmill occurred in Holland in the early eighteenth century. Excellent millwright books were published and gained wide circulation throughout Europe. The most important is considered to be "Groot Volkomen Moolenboek" written by Natrus and Polly and published in 1734, and J. Van Zyl's 1761 book "Groot Algemeen Moolen-Boek". The former contained full instructions and precise to scale drawings. Gear operation was fully explained. The teeth of the large hardwood gears driving lantern pinions were laid out by a rule-of-thumb method. Below the pitch circle the teeth were parallel, above the pitch circle the teeth are shaped with circular arcs whose centers were on the pitch circle, with a radius equal to the pitch. The Dutch writers provided more detailed and advanced working drawings on the engineering of both post and tower mills than was seen in Ramelli's 1588 "Diverse et Artificiose Machine". Very little iron was used.

In Britain there was a change from using cog and pin gears to the use of teeth with a cycloidal tooth form. This was a direct result of the scientific writings of individuals such as Leupold and Euler. Of particular interest is Leonhard Euler's book "'Mechanics, ot the Science of Motion Set Forth Analytically."

The illustration of an English mill design in figure 4-5 is of a windmill that is still open to visitors.

Figure 4-5 is a typical English (Norfolk) water-mill design 1775.

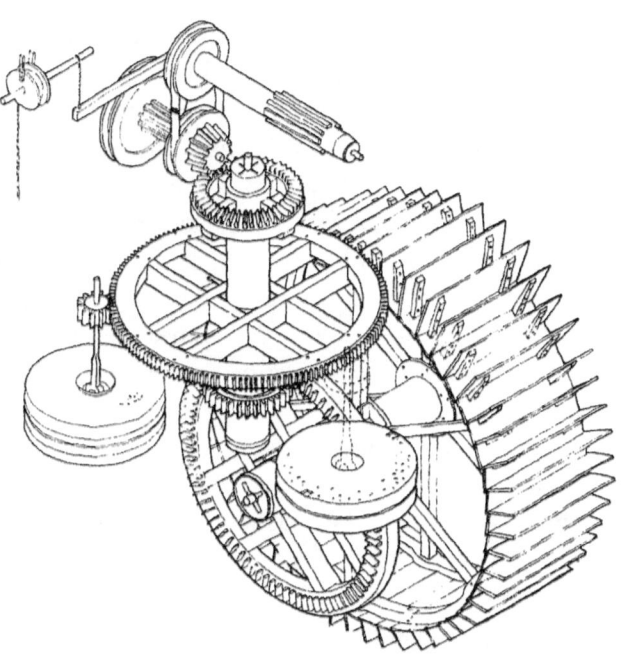

The Scottish millwright, Andrew Meikle, developed an automatic turning gear in 1750 to improve windmill efficiency. Meikle also invented the fantail that kept the sails at right angles to the wind. A small scale auxiliary windmill was mounted at right angles to the main sail. By means of a wooden worm and wheel, a pinion and rack, a speed reduction of 3,000 to 1 was obtained that drove the head until the main sails received the full force of the wind. He developed the spring sail in 1772 that counteracted sudden wind gusts. Meikle also patented in 1788, the threshing machine that could be powered by wind, horse, water and later steam, and built a factory to produce them,. This advanced machine when powered by a horse used a large gear wheel so that the horse could be harnessed underneath.

"Western wilderness gets first modern grist mill." George Washington built this mill in 1769 at the junction of the Allegheny and Monongahela rivers on Dogue Creek, on the Mount Vernon estate. They were proud of the fact that it was an overshot water wheel design ... "the very epitomy of efficient water wheel construction." The turning effort to the locally quarried buhr-stones was applied through wooden gearing. The wooden gearing and wooden machinery "are as clever and fine as that contained within any other mill in America." It was equipped with a pair of Cologne stones and at other times a pair of French buhr stones, a double geared mill.

Gear Technology: 1780 is considered by some as the beginning of the industrial revolution, others claim it had begun in the 16th century. The French introduced the term in the 19th century in association with their political revolution. One theory is it was the change from human and animal power to other power sources. The actual beginning and end is impossible to define. However, it was an important period for gear development. In "History of Manufacturers in the United States 1607-1860" the author Victor S. Clark writes, "The improvement in power transmission and gearing that enabled several machines to be run from a single wheel-shaft was undoubtedly original in America, though possibly not first invented here."

In the late (1692-1761) Professor of Physics, Petrus van Musschenbroek, at Utrecht and then at Leiden University, using Galileo's bending strength formula, systematically tested the wooden teeth of windmill gears to determine their load capacity. He also built a tensile-test machine. His book "Essai de Physique" was widely used by the engineers of the day. Musschenbroek was Dutch and is also credited with the invention of the pyrometer and discovering in 1746 the principle of the Leiden jar.

John Imison was the first to introduce the term bevel gear in a "Mechanical Power" article, London, 1787. He demonstrated the analysis of bevel gears by using the cones of intersection. The brothers Aureliano and David a San Cajetano provided a theoretical study on the dynamics of differential gearing in 1793.

Machine Tools/Lathes: In 1701 the French engraver, Charles Plumier, published in France "L'art de tourneur" the first detailed description on using an iron turning lathe, and the sharpening of the cutting tool. Plumier stated that he knew of only two other men who could machine iron on a lathe. A metal-cutting lathe was built in 1701, but its builder clearly stated that it was not the first, but one of very few.

The Huygen lathe design (1680) was improved by Antoine Thiout of France in 1741 rejecting the sliding spindle and replacing it with a sliding rest in which the tool was rigidly mounted and had a cross feed adjustment. He also produced a thread cutting machine. By 1750 he had made a very important lathe improvement in providing a tool holding carriage that by means of a screw drive moved longitudinally. The sliding spindle with a set of master gears had been illustrated in Plumier's work. This improved accuracy was independent of the eye and skill of the operator. Spur gears were used instead of ropes to rotate the lead screw.

In 1740 a major machining problem was overcome with the making of an accurate indexing plate. This was accomplished by Bird's dividing engine which took into account temperature variations and the use of a vernier. Equal distances along the arc could be measured within 0.001" and estimated to 0.0003".

Jacques de Vaucanson in 1760, the French engineer and inventor who had previously built the first automatic loom known as the jacquard loom built an improved iron frame lathe with an iron bed for turning iron. This was amongst the first heavy duty industrial lathes and is the oldest heavy duty lathe to survive. It is a complete breakaway from the pole-lathe. The Conservatoire Nationale des Arts et Métieres, in Paris, contains two medium sized lathes, one built by Vaucanson. This lathe shows a high degree of mechanical skill. The lathe has a sliding tool carriage, advanced by a long lead screw, which is parallel to the axis of the work piece. The other is Senot's metal cutting lathe of 1795 which included lubricated bearings, change gears and work rests to negate the tool thrust providing a choice in screw pitch from the lead-screw. Vaucanson designed drills and similar tools, and began the practice, which was later applied to all machine tools, of having the carriage move along a prismatic metal bench. His drill and lathe were each fitted with a tool-holding carriage moved by a threaded screw. Vaucanson was an engineer with ideas well advance of the time. The oldest known formed rotary cutter for industrial use was made by Vaucanson. (p.105)

Figure 4-6 is taken from a 1741 French book; gears in place of ropes connected the lead screw rotation with that of the work. The idea of change gears had not as yet been developed.

Figure 4-6 French Screw-Cutting Lathe of 1740

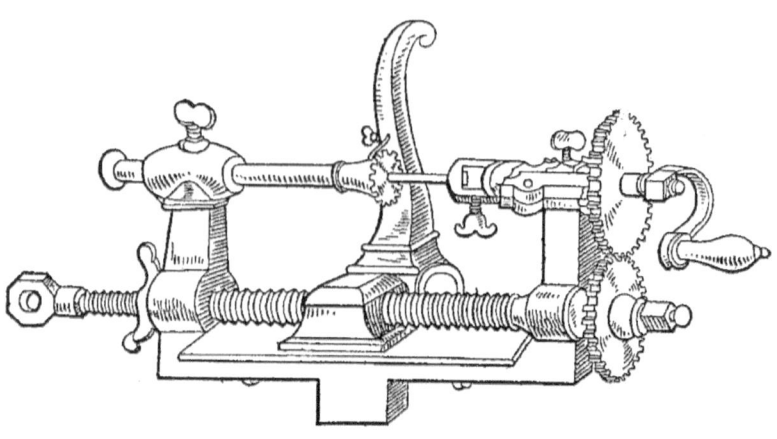

The first satisfactory screw-cutting lathe was made by the English instrument-makers John and Jesse Ramsden in 1770. The rod to be threaded was placed parallel to the guide-bench, and then rotated by gears through the driving crank of the lead screw. The tangent screw was cut on their 1777 lathe using gear wheels with numbers of teeth to obtain the fine correction of pitch. A short worm (the lead screw), cut with the best possible accuracy, had a pitch 20 threads to the inch, meshing with a large hand cranked wheel. The screw drove a very large gear wheel with a central hub carrying one end of a steel wire, which when it unwinds controls the cutting tool travel. By this means 600 turns of the hand crank mov4d the tool holder five inches. Other gears controlled the relative rotation of the tangent screw. The machine could accurately produce fine-thread screws of any length. Ramsden is said to have built a lathe with change gears in 1775.

Figure 4-7 is taken from the French "Dictionnaire des Sciences" published in 1772. The lathes were unable to cut metal as they had insufficient power and the tool holder was not rigid enough to guide the tool accurately. The work had to be rotated alternatively backwards and forwards and caught with the tool on the forward motion. The millwright tools would consist of a hammer, chisel and file.

Figure 4-7 Man Powered French Lathes of 1772

Figure 4-8 All Wood Lathe

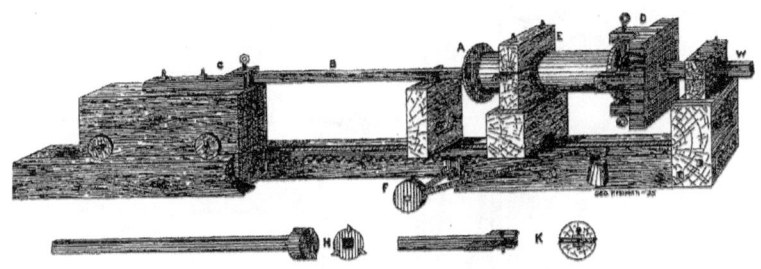

Conjectural Drawing of a Cylinder Boring Lathe of 1800. Based on a Cannon boring machine of the same period, from *Rees Cyclopaedia*, first American edition, 1802. The principle parts of the lathe are:

A—is the steam engine cylinder which is being bored out. B—the boring bar held in the clamp C—and guided by the block close to A—. D—is a simple chuck fixed on the end of shaft W—driven by a horse mill. The boring bar is pressed forward by the rack and pinion F—. In the foreground are shown the boring tools, a reamer H—, with three cutters, or as they were then called—"steelings," also a flat tool shown at K—for scraping the bore to a smooth surface.

Previously, and even into the early eighties all machine tools were built largely of wood, as shown in figure 4-8. Only the fasteners and smaller parts were of metal.

Planers: The first planing machine was built by the Frenchman Nicholas Forq in 1751. It was built to plane pump barrels used in the Marl water works that supplied the Versailles fountains. Plumier writing in 1754 on mechanical subjects described the planer as an English invention. The invention has been claimed by Spring from Aberdeen, and in England, James Fox, George Rennie, Matthew Murray, Joseph Clement, and Richard Roberts. The planer was in early use in the U.S. and may have been invented independently in that country. The earliest planer in existence was built by Roberts in 1817. (South Kensington Museum)

General: The Swiss inventor Ferdinand Berthoud wrote in 1763 on the application of machine tools to precision craftsmanship. He illustrated an automatic Fusee machine, for cutting the special screws needed for time keepers. A plate of this machine was published in 1763 as shown in figure 4-9. The invention of Le Lievre and improved by Gideon Duval it was the earliest precision machine tool design.

Figure 4-9 Fusee Machine 1763

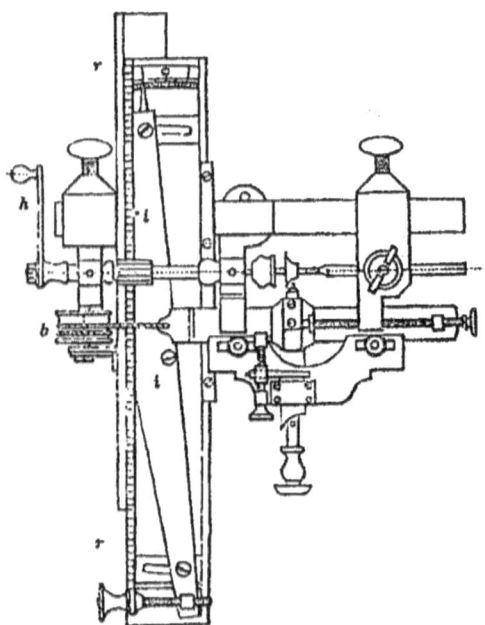

The faultless principle needed precise construction that was beyond the capabilities of that period. This early machine maintained even power as the mechanism wound down. Intricate components for clock making required this special purpose lathe, the *"fusee engine"*. The spring driven clock was not feasible before the invention of the fusee.

In 1827 Clement applied the fusee principle to maintain a constant cutting speed on the most advanced machine tool of the time, a facing lathe. A bevel gear on the shaft extension of the lead screw connected to a vertical shaft and the fusee mounted on a horizontal shaft. The gear wheel on the fusee axis has alternative rings of teeth that are engaged by a pinion that can engage with any one of the ring gears. A gut band connects the first fusee to a second fusee so that the largest diameter of one is opposite the smallest diameter of the other. The gut band follows the worm paths of the fusees from one end of their tapering barrels to the other, and by this complex method achieves a constant cutting speed.

In April, 1779 James Pickard of Birmingham, England, was granted "Patent for a mill or machine for turning, boring, milling grain, all kinds of milling or any operation quod mola facere potest rotativa motione." Pickard also held a patent on the crank as a method of applying steam engines to the turning of wheels until it expired in 1794. Watt used epicyclic gearing to circumvent the patent.

The famous Carron Iron Works near Falkirk, U.K., was installed and designed by John Smeaton in 1759. Smeaton used cycloidal gears in English windmills and in 1754 was the first to make them of iron. The gears were carefully chipped and filed into epicycloids.

The iron gears were first introduced into the Carron Iron Works. He laid down a special boring machine for the engine cylinders and cannons. (Fig. 4-10) Not patented it was copied by other engine builders. The head had inserted cutters and attempted to have a suitable movable support but failed to produce an accurate bore. It was driven through double reduction gears from a water-wheel. Smeaton and John Rennie pioneered the change from wood to iron.

Figure 4-10 Smeaton's Boring Mill 1769

When William Murdock died in 1839, his house in Handsworth was fronted by a pedestal on which was mounted an iron-toothed pinion. The inscription read: "This pinion was cast at Carron Iron Works for John Murdock, of Bellow Mill, Ayrshire, A.D. 1760, being the first tooth-gearing ever used in millwork in Great Britain."

The technical follow-up to Smeaton's iron works was the construction of the world's first large steam powered flour mill, Albion Mill, Blackfriars. It was built by Rennie from 1784-88 and the power transmission was almost wholly of iron. Samuel Smiles wrote a life of Rennie and described the installation as follows: "The whole of the wheels and shafts of the Albion Mills, were of these materials (cast and wrought iron), with the exception of the cogs in some cases, which were of hard wood, working into others of cast iron: and where the pinions were very small, they were of wrought iron. The teeth, both wooden and iron, were accurately formed by chipping and filing in the form of epicycloids." The technical advantages resulted in the adoption of iron gears for all large machinery. This was also the first time steam engine power was transmitted by iron gears. In 1791 this technically advanced mill was totally destroyed by fire created by overheated brass bearings. The production of epicyloidal gears would advance in three stages, hand built patterns for casting, machining the patterns, machining of cast or wrought gears from rough forms and blanks. Rennie also built the first screw driven vessel.

John Wilkinson was the first iron master to install Watts new steam engine. His father also from Coalbrookdale, may have been the first to follow the Darbys in the use of coke. His son was a major fan of cast iron and believed it had no limit on its uses. He is reported to have developed cast iron pipe and supplied Paris with over forty miles of his pipe in 1788. He even built a boat by bolting cast iron plates together and to the surprise of many cruised on the Severn River. He with Darby 111 proposed the new bridge to be of cast iron.

In Bersham, Wales, in 1774, considered an eccentric, Wilkinson built a boring machine for cannons and cylinders as shown in figure 4-11. His design which had previously escaped Watt and Smeaton made the boring bar more robust, ran it clear through the cylinder, and provided a fixed support at the outboard end. This was the most accurate machine built to-date and for twenty years provided Wilkinson with a virtual monopoly for cylinder boring. The hollow 15-ft boring bar still exists. The cutter head was mounted on the bar. The steam engine would now become a commercial success due to the cylinder accuracy. Wilkinson was a cast iron enthusiast, and to demonstrate the importance of iron to the industrial world he arranged to be buried in a cast iron coffin with a cast iron tombstone in 1807. He is also credited with building the first blast furnace in 1748.

Figure 4-11 Wilkinson's Boring Machine 1774

In 1799 William Murdock took charge of the Soho foundry in Birmingham, England. A sixty-four inch cylinder was to be bored. Murdock fitted the machine with a worm with triple-start and two-inch pitch, the mating worm wheel had wooden mortised teeth. This drive was very successful, it was quiet, and did not suffer from chatter or backlash. Subsequently until 1895 heavy duty machine tools adopted this worm gear drive.

Matthew Murray the English inventor and mechanical engineer established his own engineering works in Leeds in 1795. In addition to making major improvements in the design of Watt's steam engine, whose patent had expired in 1800, Murray was one of the first manufacturers of high quality machine tools for sale. He was also responsible for significant improvements to textile machinery. He introduced a screw feed in 1802.

In 1780, H. Blanc, in France, would introduce interchangeable parts manufacturing that would later become known as "The American Method."

In 1793 Sir Samuel Bentham, as an efficiency engineer in G.A. Potemkin's disorganized factory in South Russia, was the first to mass produce engineered products by machinery. Bentham was issued the famous British patent No. 1951 that stated: *"when motion is of a rotative kind, advancement (of the tool) may be provided by hand, yet regularity may be more effectually insured by the aid of mechanism. For this purpose one expedient is the connecting, for instance, by cogged wheels, of the advancing motion of the piece with the rotative motions of the tool."*

Figure 4-12 Maudslay's Screw Cutting Machine

The English engineer Henry Maudslay (1771-1831) is considered to be the greatest of the early tool builders and is known as "the father of the machine tool industry." The metal lathe with which he is best known had a cast iron bed. The lead screw tool holder was used for several pitches by utilizing change gears. He used twenty-eight change gears from 15 to 50 teeth. His greatest achievement was building metal lathes and combining the slide-rest with change gears, and a power-driven lead screw. (Both lathes are in the South Kensington Museum, London). The slide-rest was popularly known as "Maudslay's go-cart". His bench screw micrometer was accurate 0.0001 inch. (Fig. 4-12). He invented the first screw cutting machine trying for uniformity and to standardize the pitch of screws.

The French Encyclopedia of 1717 and 1772 shows a slide-rest, Bramah, Bentham and Brunel in England, Sylvanus Brown, in America, are all said to have invented this component. David Wilkinson, of Pawtucket, R.I., was granted a patent for the slide-rest in 1798, but Maudslay was the first to design and build it properly. The three great British machine-tool makers of the next generation, Richard Roberts, Joseph Whitworth, and James Nasmyth all worked under his direction. Maudslay's shop, Maudslay, Sons and Field, made the best machine tools of the period and set the standards. With Joshua Field he produced marine engines at their company plant. Maudslay is also credited with inventing the slide rule, other claim it was invented by William Oughtred.

In 1789 Samuel Slater left England for America. As a machinist he was aware of the danger in leaving due to the 1785 law and took no drawings or machinist materials. From memory he was able to redesign and build the Arkwright textile machinery. He solved the problem of spinning cotton, and was the first to manufacture cloth by use of water powered machinery. Many industry experts left Britain for America under similar circumstances.

Although da Vinci explained the use of ball bearings the first patent was issued to a Welsh carriage maker, Philip Vaughn, in 1797. They were used in the axle assembly of his carriages.

Tooth Cutting: Iron could not be fully utilized for construction without the use of machine tools. The idea of a cutting tool to precisely cut away specific portions was considered throughout the century prior to 1775 when the first large machine tools appeared. These machines needed stronger and/or more accurate gears. The London Science Museum contains a gear cutting machine circa 1672. The gear being cut was located at the upper end of a vertical spindle with the circular cutter located on a horizontal shaft behind the gear. After each successive cut the next position was located with the use of a large circular dividing plate. In a clock makers machine of 1700, the blank, was notched by a rotary cutter to form the teeth, the cutter was moved parallel to the axis between cuts. The blank was rotated and accurately positioned with the use of a perforated index plate. In Sweden, Christopher Landop produced a gear cutting machine with an index- plate and a *hypoid drive* in 1741.

James Brindley was engaged in 1755, to make the larger gear wheels for a silk mill near Congleton, Cheshire, England. One economy was contriving a method to cut all his gear teeth by machinery. Previously they had been hand cut. The work that had previously been accomplished in fourteen days now took one day.

In 1765, Henry Hindley, clock maker in York, England, was credited with designing worm gearing with both elements throated, i.e., globoidal, also known as the Hindley worm. The gears were used on a dividing machine. The 360 tooth gear was 13 inches in diameter, with a1/16th face, and 1º helix angle. John Smeaton of whom we have previously written provides this description: "The threads of this screw were not formed upon a cylindrical surface, but upon a solid, whose sides were terminated by arches of circles, the screw and wheel, being ground together as an optic glass to its tool, produced that degree of smoothness in its motion that I observed, and lastly, that the wheel was cut from the dividing plate." This hand powered gear cutter used differential indexing for cutting pinions and racks. Professor Mac Cord, describing how the gears were manufactured wrote that they were very easily made with a tool whose cutting point had the contour of the wheel's tooth "… is so clamped to a disc that its upper surface lies in the meridian of the plane of the worm, and both the disc and the worm blank are driven by intermediate gearing at their proper relative velocities."

Fig. 4-13 Hindley's Gear Cutter 1741

The indexing plate had reamed holes to engage a tapered pin. This important feature was far ahead of its time and would not be utilized until adopted for milling machines one hundred and fifty years later. The plate had a hypoid gear drive.

Dauthiau claims to have invented the differential in 1735, fifteen years earlier Joseph Williamson had made similar claims. We now know the differential has been is use at least as early as 400 B.C. and, in the 19th century, others such as Richard Roberts in 1833 would make the same claim. Roberts did design a steam carriage with a differential drive to the wheels, its first appearance in a road vehicle. Baldewins's clock contained a differential in 1575, as did Salomon de Caous's pump steam engine in 1615.

The French Duc de Chaulnes (1714-1769), who introduced the use of microscopes with cross hairs, and using Bird's method for making indexing plates, experimented with a mechanism for graduating the circle mechanically. In England in 1763 Jesse Ramsden developed a machine for

A Gear Chronology

the same purpose. Accurately cut instrument gears and screws were now capable of being made for use in observatories etc. Their second dividing engine was forty six inch diameter and is in the U.S. National Museum with the equipment with which it was made. (U.S. National Museum Cat. # 2125518 – The device for cutting the worm #215519)

Milling cutters, ground to an approximate form for conjugate action, replaced rotary file cutters due to the work of Samuel Rehé in England. His gear cutters, while not as accurate for clock making as Hindley, provided a transition from clock maker to machinist's gear. Rehé also built a cutter grinder, and provided attachments for cutting internal gears (circa 1780). He ground his cutters on a cylindrical grinder using a mixture of emery and oil which provided very precise grinding of the cutting edges. The elliptical cutting edge was inclined to give an approximation of the correct form. The cutters are closer to milling cutters than rotary files. His wheel cutting engine was another transition from clock making tool to a gear cutting machine. It had a heavier more massive frame than previous machines.

The oldest known rotary file cutter was made in France by de Vaucanson in 1782. The teeth were cut by chiseling. The octagonal bore appears to have been broached. The cutter was in the possession of the tool builder Brown and Sharpe.

One section of Leupold's "Theatrum Machinum" (ref. p.89), illustrates a machine that provided the first evidence of cutting gear teeth by broaching. It was also the first machine to give flat bases to the spaces between the teeth.

In Madrid an early successful gear cutting machine was built by Manuel Gutierrez as shown in the figure 4-14.

Figure 4-14 Manuel Guiterrez Gear Cutter 1789

Gear Applications: Water wheels and pumps were installed at London Bridge in 1751. The wooden axle was nineteen feet long and three feet in diameter. Two forty-four tooth spur gears were mounted to the axle to drive the forty spindle lantern wheels keyed to the cast iron crankshafts as shown in figure 4-15.

Figure 4-15 London Bridge Water-Wheel

In 1775 Sir Richard Arkwright the English inventor and industrialist used iron bevel gears in the cotton mills at Cromford and Belper. He was innovative and his mills changed from horse to water and then to steam power. His inventions considerably reduced the manpower needed in the cotton mills causing a mob to destroy his main mill.

The 1780 illustration figure 4-16 is of a crude worm gear driven jack with a pinion using rollers in place of teeth. It is indicative of the then current gear technology.

Figure 4-16 Worm Geared Jack

CHAPTER 5

Gear Development and Growth In The Nineteenth Century

Introduction-
Power – Materials -- Testing – Measurement-- Machine Tools – Grinding -- Gear Manufacture – Cutters – Tooth Forms -Transport -- Lubrication

Introduction: The years between 1750 and 1830 were significant in the development of iron metallurgy and how it could be worked. This advance was combined with the progress in the use of steam engines. These engines required precision parts and new machine tool designs to produce them and as one requirement would lead to another so industries would develop. Prior to the 19th century the necessary capital for the required heavy machine tools was unavailable even if the technology existed, neither were the production quantities to make machine shops profitable.

The beginning of the century provided major difficulties for manufacturers in Britain and the U.S. Organized gangs – the Luddites – smashed machinery that they believed would take away their livelihoods. Parliament in 1812 decreed that destroying or sabotaging machinery was punishable by death, seventeen leaders were summarily hanged. In the U.S. industry had come to a virtual halt with the passing by Congress of the "Embargo Acts (1807-1809)" that prohibited all American exports.

However, before its repeal Jefferson encouraged the expansion of domestic industries and this growth would continue throughout the century and beyond. The increase in manufacturers was so rapid that by 1868 Bishop's "History of American Manufacturers" published in Philadelphia would require three volumes.

The 19th century would see a demand for increased power, improved materials, heavier machine tools and high quality gears. In 1900, 92 percent of the energy used in the world was supplied by coal. This was a remarkable change brought about over the twenty years 1850 to 1870. In 1850 half of ocean going ships and half of the world's railway tracks were British. Five times as much iron was smelted in Britain as in the U.S. and ten times as much as Germany. The power from Britain's steam engines was equal to half that of the whole of Europe. The U.S., Germany, France, Switzerland and Belgium had joined Britain by 1870 in having self-sustained economic growth and the U.S. would be far ahead of the others in steam power. In 1897 the U.S. would lead the world in production of steel accounting for 34.58 percent of the world total. This growth was accelerated by the demand for improve agricultural methods, bicycles, rail and marine transport, and steam engines.

Prior to the arrival of machine printed books, in the latter part of the 16th century, few technical records are available. Earlier, most drawings had been mere visualizations of ideas that could then be turned into models by skilled artisans. These early inventors did not make finished drawings and patents could be obtained by submitting wooden models, and/or by a description in legal language. In the 19th century drawings and technical books would become commonplace. Early in the century several engineering books were published by German, French and English writers that contributed to the advancement of machine engineering. Included amongst these books was Gaspard Monge's "Elements of Machines", Lanz and Bétancourt's "Essai sur la composition des machines" written in 1808, Jean Nicholas Pierre Hâchette's treatise "Traité élémentaire des machines" written in 1811 is considered to be the foundation for the systemization of mechanical engineering. Tycho Borgis wrote "Tarité complet de mécanique" in 1818. The French mathematician Baron Augustin Louis Cauchy published the influential book "Cours d'analyse" in 1821. Cauchy contributed important work on elasticity, partial differential equations, and the theory of functions of a complex variable. Charles Babbage, the English mathematician and pioneer of the modern computer, with Andrew Ure an

English chemist, wrote" Economy of Machinery and Manufactures" in 1825 and "Philosophy of Manufactures" in 1830, describing the factory as the physical embodiment of mathematical principles and a self-acting machine.

The first published mechanical works useful to the machine shop, "Nicholson's Operative Mechanic", was printed in London, 1825, with an American edition published in Philadelphia in 1826. Also of advantage to the growing industry was Dr. Alexander Jameson's "Mechanical Dictionary." There had been no trade magazines until John Williams, an immigrant from Ireland, changed the name of his "Hardware Man's Newspaper" to "The Iron Age" in 1859: "…to keep in view prominently the manufacturing capacity of the country, and to apprise our readers timely of everything important to the iron trade, whether in this country or Europe."

Charles Holtzapffel, a German immigrant to England, and his son John Jacob, produced a massive work in five volumes; "Turning and Mechanical Manipulation" in 1843. A sixth was planned but never published. The books became the guide for metal turning. This was the fullest and best detailed description in the state of the art in machine tools to that period. Both Richard Roberts and Whitworth had been employed by Holltzappfel. The books have been described as "possibly the greatest work in the English language dealing with the lathe and its accessory apparatus." Holtzapffel also designed a self-centering chuck in 1811. The first American text-book devoted to mechanics, "Elements of Analytical Mechanics," was written by a West Point professor, W.H.C. Bartlett, and used at the Military Academy until 1887.

In 1866 there were only six engineering colleges of established reputation in the U.S. In England, Whitworth saw the value in technical education and in the late 1830's supported the Mechanics Institutes and Manchester School of Design leading to the Whitworth Scholarships in 1868-9. The total number of graduates over the thirty-one years prior to 1866 was only three hundred. By 1870 the number of colleges had grown to twenty-one and 866 engineers had graduated. Ref: Wickenden Report of Investigation of Engineering Education (1923-29 Vol. 1, page 542). After 1850 the former mechanics would be known as engineers. This increased demand for engineers resulted from the manufacturing of precision parts which required blueprints, tolerances and calculations. Such was the growth in engineers that by 1856, the Scottish writer Samuel Smiles, could write an industrial biography titled "Lives of the Engineers".

He had previously written "a Life of George Stephenson." In Britain the technical schools were negatively affected when the unions were able to pressure Parliament to pass in 1889 "The Technical Instruction Act" that prohibited the teaching of a specific trade.

The increasing use of power for transportation and manufacturing created an increased demand for technical training. The universities for the first fifty years of the century would have no interest in providing an engineering education beyond the existing apprenticeship level. On the occasion of the "Great Exhibition" in 1851 it was declared by some that industry from now on would not depend on local advantages by on a competition of intellects. Negligible attention was paid to science at Oxford and Cambridge. There was virtually nothing in Britain to compare with the technical colleges of France and Germany.

The U.S. made significant industrial advances in the 19th century. Willis writing in 1841 noted that Britain transmitted from the prime mover (steam engine, water-wheel or turbine) to various factory machines by long shafts and toothed gear wheels and in America by large belts. Factories and machine shops in which a central steam engine powered the majority of machines through a system of belt transmissions would exist well into the 1960's.

The demand for gears also grew at an amazingly fast pace as did their production as can be seen from the following example. Johann Renk established a gear factory in Augsburg, Germany, in 1873. Twenty years later the plant was producing 12,000 gears per annum.

Power Developments – Horse Power- Water/Wind-Steam-Electric-Gas

Horse Power: The use of horse powered machinery remained low for three quarters of the century but in the last quarter of the 18th century real growth took place on the farms. Horsepower on treadmills might involve as many as three horses. Powered sweeps used as many as seven horses hitched to poles similar to a ship's capstan, transmitting power through a train of gears to provide rotary power. Horse-wheel houses were built in great numbers; more than 1300 have been identified in the British Isles (Agricultural History Review, vol., 24, 1976), manufactured through the 18th and 19th centuries. Early in the 19th century the universal use of cast iron allowed for a more efficient design. The central shaft was usually short and carried a small diameter cast iron crown wheel. The first gear wheel would engage a smaller bevel gear that connected to the farm machinery by a universal joint. On other occasions the horse-wheel frame would include a gear train. Many of these horse powered engines were exported and the U.S. manufactured vast numbers to their own design sometimes capable of accommodating as many as twelve horses. Some mistakenly believed even in the 20th century that the horse would always be required for power and they remained in use for transport and agriculture long after they were proved to be uneconomical.

Wind Power: Since the earliest times until the present day wind and water have been used to generate power. Gears were a necessary component of the mill's design.

An impractical wind turbine had been constructed with the intention of uprooting trees in 1680. Twelve horizontal shafts each with a pinion and gear were to rotate a winch type mechanism. Even though the 19th century saw a movement away from wood to iron and steel, wind and water mills still tended to use wooden gears. Holtzapffel detailed the properties of several different woods in his Volume1 on Machining in 1843 and stated the principal woods used for gear teeth were from the crab tree, hornbeam, and locust. Darle Dudley in his book "The Evolution of the Gear Art" describes a visit in 1967 to the Thompson Manufacturing Company, Lancaster, New Hampshire, and their supplying "tens of thousands per year" of maple gear teeth for mills in nearly every state east of the Rocky Mountains.

Prior to 1850 Nathaniel Dominy built three windmills, one of which, the Hook Windmill, was built at East Hampton, New York, in 1806 and is a rare surviving example of his craftsmanship. Its sophisticated gearing was made of wood. During 1850-60 an Illinois mill was built with German, Dutch and Swedish workers. The German craftsman Louis Blackhaus used hickory and maple gearing. Known as the Fabyan Mill it is the best example of a Dutch windmill design that can be seen in the U.S. today.

A major understanding of the design of wind and water powered machines arrived with the publication in 1849 of Professor Julius Weisbach's "The Mechanics of Machinery and Engineering." The two volumes were profusely illustrated with eight hundred and thirteen wood engravings. Weisbach was Professor of Mechanics and Applied Mathematics at the Royal Mining Academy of Freiberg. The power is transmitted by a bevel gear from the wind shaft to the vertical shaft. On the regulation of wind power he wrote "As the wind varies in intensity as well as in direction, when the work to be done is a constant resistance, unless some means of regulating the power be applied, the motion of the machinery would not be uniform.

Weisbach suggested the use of a friction strap or varying the extent of sail, or quantity of exposed cloth as a solution. Self-adjusting sails had been invented in 1817 by a Sir. William Cubitt. He used pinion and rack gear sets as shown in figure 5-4.

Figure 5-1 Windmill Gearing Circa 1850

Figure 5-2 Hook Windmill Gears 1806

Figure 5-3 Weisbach's Illustration of a Smockmill

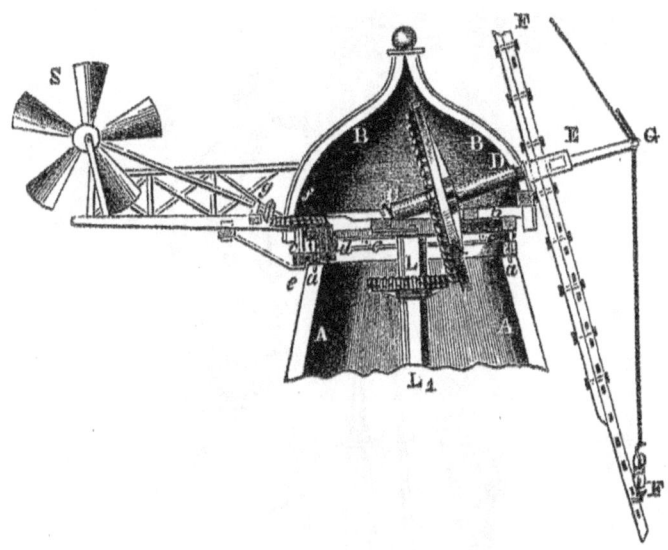

Figure 5-4 Method for Self Adjusting Wind Sails

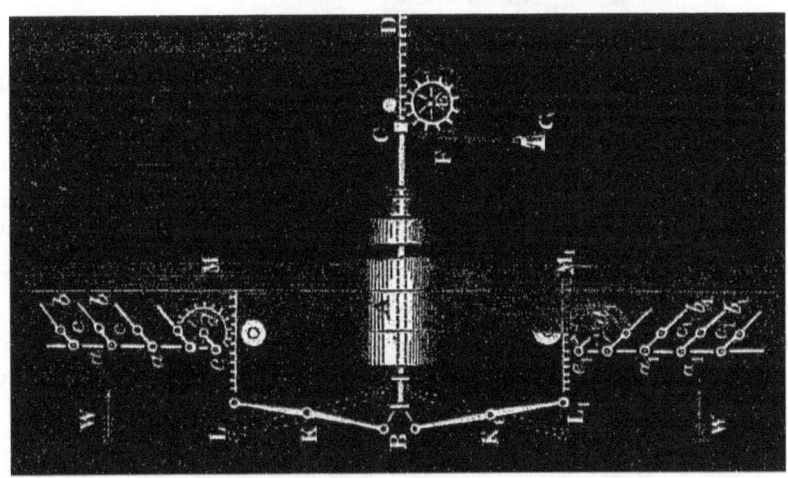

Figure 5-5 illustrates a mill built by Goodhue Wind Engine Company and is taken from an advertisement in "Farm Implement", March 22nd, 1894. By the end of the 1890's many of the early single gear mills had been replaced by double-gear mills which used two gears and two pinions for extra strength and smoother action. In 1899 Red Star Windmills was advertising "The Only Mill on the market with Steel Main Gear."

Figure 5-5 Goodhue Windmill

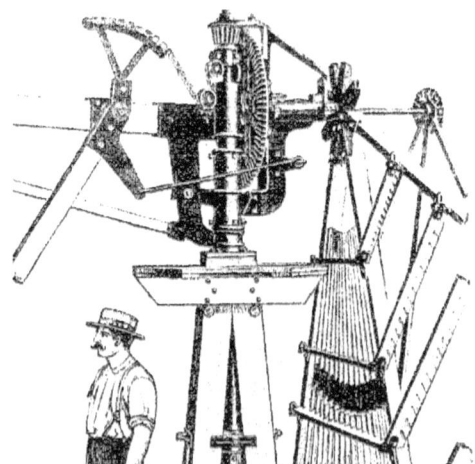

Water Power: The English translator Prof. L. Gordon wrote in the advertisement for Weisbach's books "...the only Theoretical Treatise on Water Power...it is the first publication in which a systematic attempt is made to familiarize English Machinists with the application of exact reasoning in developing the theory of machines..." Every detail of the various water wheel designs was included:. "Overshot wheels have been constructed for falls varying from eight to 50 feet, and sometimes even up to 64 feet in height... up to 50 cubic feet of water per second"

Weisbach provided the principles and rules to dimension the components based on the weight and power of the wheel. The average power of the U.S. water-wheels reported in the 1870 census was 42.2 hp and in the 1880 census 45.85 hp. Numerous formulae were provided covering such items as the resistance of the axle to transverse strain. He stated that the prime mover is often a toothed wheel, forming the periphery of an outside crown as shown in figure 5-6. "The pinion should be rather below than above the level of the axle, and on the side on which the water is." It is truly amazing that prior to 1849 such large gears could be produced without ovality.

Figure 5-6 Weisbach's Illustration on The Construction of Water Wheels

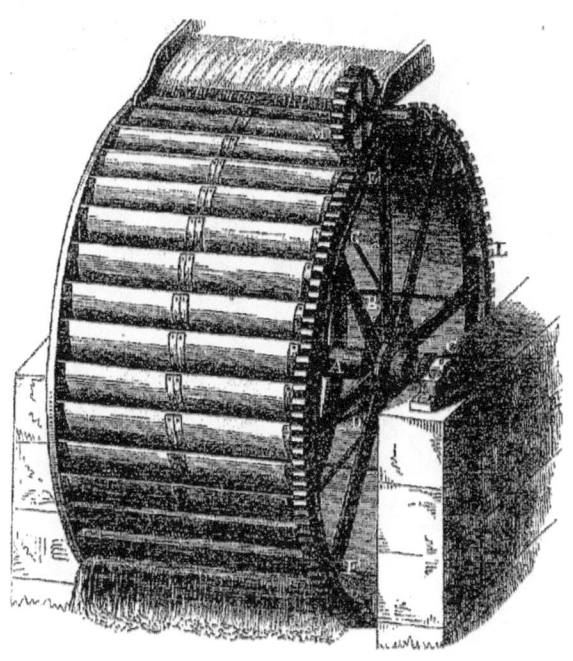

Weisbach also reports on the development in hydro turbines. Patents such as that obtained by Manouri d' Extot in 1813 and Whitelaw's 1843 patent "Scottish Turbine" as shown in figure 5-7. Other turbines by Combe, Cadiat, Fourneyron and Redtenbacher indicate a much wider use than was formally believed. Most required the use of a right angled gear set. The French engineer Benoit Fourneyron improved the efficiency by twenty percent. The 1827 prototype outward-flow radial turbine developed 5hp with an efficiency of 65 percent. By 1833 his turbines developed 50hp with 73 percent efficiency. In 1895 his turbines were installed at Niagara Falls. Due to poor part load performance they would be replaced with Francis Turbines.

Figure 5-7 Whitelaw's 1843 Scottish Turbine

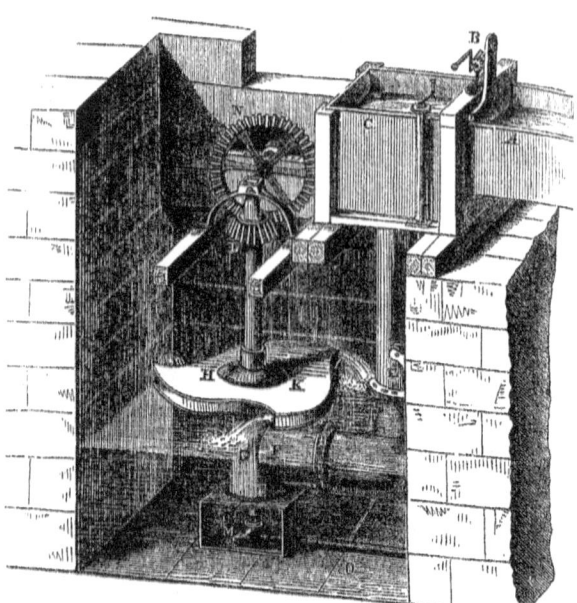

Steam Power: The first steam engine installed on the European continent had taken place in 1722 at Königsberg, Russia. By the first part of the 19th century steam power would be extensively used for stationary power, transport and ships, the golden age of the steam engine. The reliability of these early steam engines can be seen from the history of an 1810 steam engine that powered a winch in Rutherglen, Scotland, and would not be decommissioned until 1915. Until 1800 none had developed more than fifty horsepower but in the next decade Trevithick in Britain and Oliver Evans in the U.S. had built engines generating a hundred horsepower. By the end of the century powers of 10,000 horsepower were not unusual. The 1870 U.S. census gave the average steam engine horsepower as 112.72 and by the 1880 census 118.79. To supply the engines ln the year 1800 coal production of ten million tons had increased to sixty million tons in 1850.

The use of steam engines for factories spread rapidly In 1838 the Secretary of the Treasury reported to Congress that 3,010 were in use, 800 on steamboats, 350 in locomotives, and 1,860 in factories. The steam engine allowed for manufacturing factories that could be expanded, and located anywhere. The steam engine also increased the need for precision and over the next seventy years accelerated the development of new machine tools. In 1860 only thirty percent of the world's shipping was steam powered but by 1894 the figure had risen to eighty percent. Another indication of the rapid growth can be seen by the installations in the U.S. increasing from 3,500,000 hp in 1860 to 16,940,000 hp in 1895, which probably represented fifteen percent of the world's total at that time.

In 1801 Evans built his first stationary high-pressure steam engine to power a rotary crusher. Over the next several years his steam engines competed successfully with the low-pressure Watt design. Evans received a patent in 1804 that was extended in 1815. His engines powered saw mills, flour mills, boats and boring machines. In 1854 Evans built a steam engine powered stern wheel paddle boat for use on the Delaware and Schuykill rivers. He more than any other was responsible for the wide use of the steam engine. He dredged the Philadelphia docks by driving his steam driven dredge under its own power from his workshop. Another of his engines drove a screw mill that could break twelve tons of plaster in twenty-four hours. His engines were smaller and more compact than low-pressure engines.

Farm machines always required power and steam engines were used for driving belts as early as 1820. Stationary steam engines were used on the farms prior to their use on railroads. Many farms in the South had such engines installed from 1807 on. 585 farm steam engines were in use In the U.S. by 1838 driving saw mills, sugar mills, grist mills, threshing machines and cotton gins. In 1849 mobile steam engines mounted on wheels and moved by horses were introduced. During the 1870's they became self-propelled and powerful enough for plowing. Before the end of the century there was enough steam power on the farms to equal the power of seven million horses. The engines developed over 110 horsepower, weighed as much as twenty tons, and the Belt and Holt engines could out pull forty mules.

Both Murdock and Nasmyth developed steam engines. The electric motor was still decades away and they both worked with adapting their small vacuum and steam motors. On his visit to the Soho foundry in 1830 Nasmyth wrote "There I observed Murdoch's admirable system of transmitting power from one central engine to other small vacuum or atmospheric engines attached to the individual machines they are set to work. The power was communicated by pipes led from the central air or exhaust pump to small vacuum or atmospheric engines, devoted to the driving of each separate machine, thus doing away with all shafting and leather belts, the required speed being kept up or modified at pleasure without in any way interfering with the other machines. This vacuum method of transmitting power dates from the time of Papin; but until it received the masterly touch of Murdoch it remained a dead contrivance for more than a century."

Murdoch had taken charge of the Soho Foundry in 1798 and made major changes to the machines and their operation. Line shafting remained in wide usage. Nasmyth designed a self-aligning bearing as with the available bearings long line shafts could not keep their alignment. Following his visit to Soho Nasmyth used small steam engines to power individual machine tools at the Bridgewater Foundry, England.

Steam Vessels: The application of steam power for boat propulsion has a long history going back at least to Blasco de Garay in 1543, but there was no practical steam vessel until the 19th century. The first steamboat, the single paddle wheel "Charlotte Dundee", was built by William Symington, and in 1801 was put to use on the Forth and Clyde canal in Scotland. In 1804, in the U.S., the American engineer Robert Fulton, having previously built a mill for processing marble, launched a steam driven vessel on the Hudson River in 1807. The "Clermont" was the first successful steam boat. It has been described as America's first great invention. It was a 142 feet long, 14 feet wide "side-wheeler."

The best engines of the time were manufactured in Britain by Matthew Boulton and James Watt. With a twenty-four inch piston and coincidentally twenty-four horsepower they were powering British factories in the industrial revolution. Britain had placed an embargo on selling this technology to the U.S. However, through the intervention of the U.S. Minister in Britain, James Monroe, an engine was waiting for Fulton when he returned to the States in 1807. When the boat stopped, the B & W engine was kept running by disengaging the gear train. Frequent failures occurred with the cast iron shafts on which the fifteen foot water wheels were mounted.

Figure 5-8 Fulton's "Clermont"

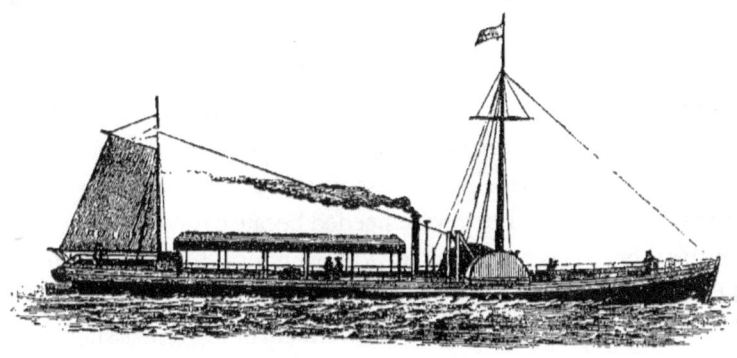

Colonel John Stevens built a steamboat with a twin screw driven by iron gears that doubled the boat speed. (Figure 5-9)

Figure 5-9 Stevens Steam Powered Vessel

Steamboat patents were filed by several inventors, including James Fitch, Nicholas Roosevelt. A boat designed by James Rumsey was driven by an inefficient water jet and demonstrated on the Potomac in 1787.. The jet itself absorbed most of the thrust instead of the boat's propulsion. A similar design had been advanced earlier by George Washington and Benjamin Franklin.

The Harrison patent simplified the steering of steam powered vessels. The engine was driven through worm gears. Large worm gears were also used for Barring engines. These auxiliary drives would turn the engine over at slow revolutions for maintenance, adjustment, or starting purposes. Also In England William Frederick Howe invented a link-motion reversing gear for steam engines in 1842.

Steam Turbines: Making direct use of water, wind or steam jets had been contemplated since Heron. A Hungarian, Baron Kempelen, gave consideration to a steam turbine which thirty years later would be further developed by Trevithick. The successful steam turbines were developed from the water-turbine. They were needed because towards the end of the 18th century there was an urgent need for engines to drive dynamos at a speed far beyond the range of the steam engine.

In 1884 the first patents for a steam-turbine were issued to C.A. Parsons, younger son of the Earl of Rosse. The first steam turbine ran at 18,000 rpm. When the turbines were used for ship propulsion maximum efficiency called for higher speeds that were totally unsuitable for the screws, reduction gearing was therefore required. The selection resulted in double reduction units for merchant ships with screw revolutions of 100 rpm or less. Marine progress was rapid and by 1897, at the British Jubilee Naval Review, the Turbinia with three turbines developing 2,000 hp ran at the then unheard of speed of 34.5 knots. Within a decade power units of 70,000 hp were being installed. Parsons also purchased the S.S. Vespasian, removed the triple expansion steam engines and substituted two 500 hp at 1500 rpm steam turbines connected to the propeller at 75 rpm through herringbone gears. The pinions were cut from solid on a shaft of soft grade chrome-nickel steel. The two twenty tooth pinions meshed with 398 tooth rolled-steel gear rings mounted on a cast iron spider, all enclosed in a gear case and lubricated by oil jets. Because of the pinion widths bearings were placed between the left and right-hand teeth. The involute form teeth had a 4 diametral pitch, 23° spiral angle, 20° pressure angles, and a thirty-four inch face width allowing ten inches for bearings. The gears ran for a year covering over 20,000 miles with negligible wear (0.002 inch at pitch line), and at an efficiency including bearing losses of 98 percent. There was an all round fuel saving of twenty percent over the original coal fired engines.

Carl Gustav Patrik de Laval, the Swedish engineer, in 1890 built a single disc steam-turbine of a different design to Parson that rotated at 30,000 rpm, requiring the engine to be geared down. In essence it was a steam driven windmill, the jets impinging on vanes located around a wheel rim. Of relatively low power until Curtis in the U.S. built additional rows of vanes.

Steam Vehicles: In 1801 Richard Trevithick built the first practical steam powered passenger vehicle in Coalbrookdale, England, considered by some to be the first automobile. The vehicle (figure 5-10) weighed ten tons and was capable of ten miles an hour on the flat. The drive wheels were ten feet in diameter. Large pinions were driven by the connecting rod, meshing with the drive wheel gears. Trevithick had previously built several tricycles for transporting light goods. He used higher steam pressures than had been used previously (50lb/sq. ins). The boiler had a fly wheel though which the spur gears drove the wheels on one side only.

Figure 5-10 Trevithick's Passenger Vehicle

Figure 5-11 Pen-y-Darren Locomotive 1804'

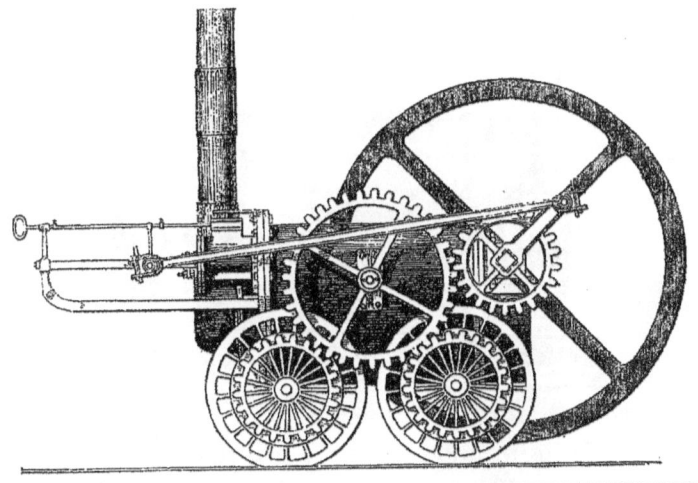

In 1804 Trevithick built a second locomotive that was capable of hauling up to fifteen tons of iron at the Pen-y-Daren ironworks. (Figure 5-11)

The forerunner of power driven passenger road vehicles began in Britain with gear driven steam powered omnibuses. In 1831, due to pressure from Britain's horse driven coach drivers, legislation restricted further development of steam carriages. Two major accidents resulted in widespread public reaction against vehicles and a seven mph speed limit was imposed. Other numerous hostile acts were also passed against steam vehicle usage. Gears were a major component of the transmission design. Several steam - powered land carriages were built and driven in the U.S. prior to 1850. In 1851 the first bona-fide manufacturer, the American Steam Carriage Company, N.Y., was founded. Prior to this time William H. James had built several such vehicles. One of his vehicles with a two cylinder reciprocating engine could be seen on the New York streets in 1829. Another of his creations carried twelve passengers at fifteen miles per hour. His vehicles had no differential and literally had to skid around corners.

Figure 5-12 Blenkinsop's Rack Locomotive 1812

From his desire to improve locomotive traction John Blenkinsop, Leeds, patented in 1808 a rack work rail and cogged wheel. (Figure 5-12)

A Gear Chronology

Mathew Murray, the English inventor and mechanical engineer, built the first commercial steam locomotive and rack railway in 1812, which was in general use for several years. Driven by a geared driving wheel for maximum traction, and operated by the Middleton Colliery Railway it hauled coal three and a half miles to the city of Leeds.

Rack railways are still in use in many parts of the world. One such rack railway was built to take passengers to the top of Mount Washington. The first engine "Hero" was received in 1866. The steepest part has a 37.41 percent grade, the front passengers are fourteen feet above the rear passengers. The design was based on a model cogwheel engine built by the inventor Herrick Aiken. Beneath each steam engine are two heavy cog gears, one on each axle. They are driven by spur gears larger than the locomotive wheels. The rack consists of thick bolts equally spaced on four inch centers.

Dr. Joseph Buchanan constructed a wheeled vehicle powered by his light weight capillary steam engine. Buchanan also anticipated it would power a flying machine. The vehicle was driven three or four miles through Louisville, Ky. All early steam vehicles were called locomotives regardless of their operation on rail or road.

September 27th 1823 is the date given as the beginning of the Railway Age. Although the idea of rail transport can be seen as far back as 1530 this date was the start of the first public steam line, the Stockton and Darlington Railway In England in 1813, the first locomotive Puffing Billy was built based William Hedley's the test proven principles. (Figure 5-13) The improvement over its predecessors was due to the one sided single crank connected to gear wheels the teeth were made with sloping sides. In 1814 Hedley introduced a significant design change by gearing each wheel. As shown in figure 5-14 the locomotive was also the first to use flanged wheels.

Figure 5-13 Puffing Billy

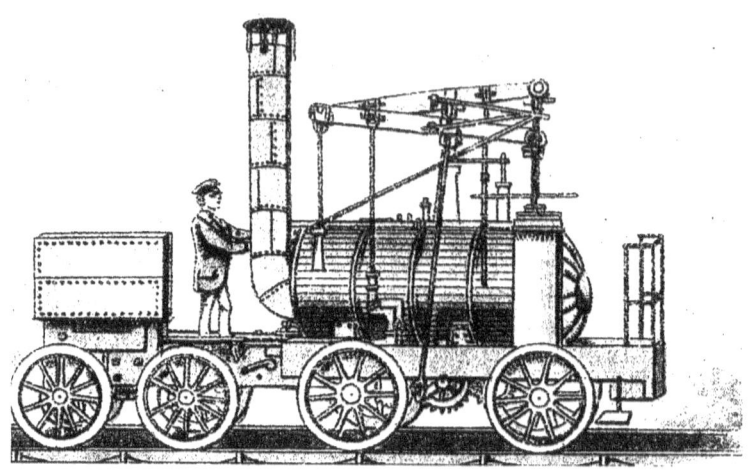

Figure 5-14 Hedley's Locomotive of 1814 each wheel was geared.

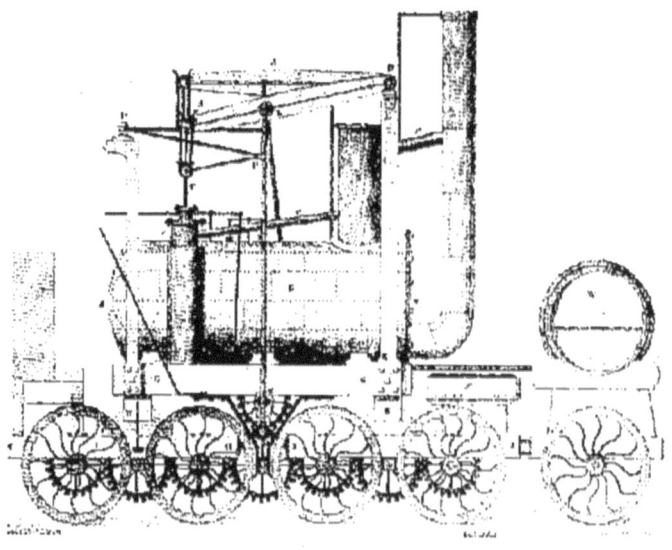

Peter Cooper raced Tom Thumb, the first locomotive made in his Canton Iron Works, Baltimore, against a horse-drawn rail car in 1829. He chose a combination gear and belt drive transmission to reduce engine speed to a practical track speed. The race was lost because the belt slipped from the pulley. The race proved the practicality of mechanical drives.

Electric Power: In 1796 the Italian physicist Alessandro Volta demonstrated a battery. This was the first device to supply a sustained and steady current flow. Electricity was produced by the contact of different metals in a moist environment. The battery led to the development of the complete range of applications powered by electricity.

The 19th Century was important for the invention, introduction and application of electric motors. Before Michael Faraday, the English scientist, generally considered the greatest of all experimental physicists, generated electricity from magnetism electricity was known either as a natural phenomenon or a laboratory experiment. Faraday more than any other was responsible for the electric culture of today. He was living in the steam age based on coal he alone believed in the future of electricity.. His first success was discovering the principle of the dynamo. By inducing a current from magnetism. In 1831 he led the way for the development of the generator and modern power station. dynamo and generation of electricity by machine. Faraday in1821 assembled the first electric motor in which a suspended electrified wire circled the pole of a magnet. Three years later he discovered benzene in tar derived from coal, the first liquid fuel that could be used for internal combustion engines. His life work "Experimental Researches on Electricity" was published over a forty year period. During 1821 he used an electric current to produce mechanical motion. The basic principle for the induction heating process is electromagnetic induction that Faraday discovered in 1831. It would not be until the early thirties that induction heating would be recognized as a production tool for heat treating. Joseph Henry, American physicist, first secretary of the Smithsonian, discovered electric induction independent of Faraday, and in 1828 built the first electro-magnetic motor, the "henry" inductance unit being named for him. His motor provided uniform motion at a rate of seventy-five vibrations per minute. Using a large battery and very weak acid the motion could continue for an extended time period. Faraday made numerous discoveries some of which would lead to radio, radar and television.

In France André Marie Ampère presented in 1822 a theoretical basis that would later be used for the invention of generators and electric motors. Ampére, a French physicist, whose name is given to the basic SI unit for electric current, in his 1830 essay "Essai sur la Philosophie des Sciences" used the name "Cinematique (Kinematics)" derived from the Greek for his studies on motion.

With the publication of his book, "The Analytical Theory of Heat" in 1822," a Frenchman Joseph Fourier, made a major engineering contribution. Fourier conceived the concept of flux, heat transfer, and that engineering equations must be dimensionally equal. The English physicist James Prescott Joule In 1834 demonstrated the nature of force. Joule determined the mechanical equivalent of heat by four different procedures in 1841 and, the First Law of Thermodynamics. In the 1846 issue of "The Philosophical Magazine" Joule established the precise relationship between heat and work, crucial to the design of every heat-engine. The unit of energy, the joule, was named in his honor. His tombstone is inscribed with the number 722.55, which was his approximation of the universally accepted 778 foot-pounds per British Thermal Units.

The English scientist William Sturgeon's rotary motor followed in 1832. In 1825 he had constructed the first practical electromagnet, and in 1836 the first moving coil galvanometer. Also in 1836 Sturgeon compiled "Annals of Electricity" the first British journal on the subject of electricity.

Mechanical generation of electricity was demonstrated in Paris by Hippolyte Pixii in 1832. The generator was fitted with a commutator to provide direct current. In '34, Jacobi, with the support of the Emperor Nicholas First of Russia used his rotary electric motor to drive a twenty-eight foot long boat with fourteen passengers at three miles an hour through a right angle gear arrangement.

Davenport's patent incorporated a bevel gear set as shown in figure 5-15. In 1840 he used his motor to print a news sheet under the title "Electro Magnet and Mechanics' Intelligence". Another of his achievements was the building of a small circular railway in Springfield, Mass. Numerous patented designs followed such as: Thomas Davenport (U.S. Pat. No. 132) in 1837. Walkley in '38 (U.S. Pat. No: 809), Stimson in '38 (U.S. Pat. No. 910), Cook in '40 (U.S. Pat. No. 1,735), Elias in Holland in 1842, Lillie in 1850 (U.S. Pat. No. 7,287), Dr. Charles Page's motor of '39 was improved and patented in '54 (U.S. Pat. No. 10,480). This latter motor powered a train from Washington to Badensburg at nineteen miles an hour. All these motors used voltaic batteries and their high cost precluded popular usage. Dr. Page, Salem, Mass., is also credited with building the first electric generator in 1838.

Figure 5-15 Davenport Electric Motor

Another innovative invention and adaptation of the new electricity took place in 1839 when Robert Anderson, Aberdeen, Scotland, built an electric powered vehicle. The Belgium electrical engineer, Zénobe Thèophile Gramme, constructed the first commercially useful direct-current dynamo in 1871 that proved to be an immediate success. Electric power could not be properly utilized until Gramme's invention. In the Vienna Electrical Exhibition, electricity as the prime mover for machine tools was first demonstrated. The following year Gramme installed an electric motor to drive the line shafting at his Paris plant. The French engineer Hippolyte Fontaine was in charge of Gramme's generator display. A workman erroneously connected two dynamos, and when Fontaine started the first one he noticed that the second dynamo was turning in the opposite direction. From this mistake two important principles were established: first electric power could be transmitted over considerable distance, secondly a generator working in reverse became a motor transforming electrical energy into mechanical energy. This direct current electric generator was the first to be used commercially, for electric plating as well as lighting.

The Electrical Exhibition, held in Paris in 1881, demonstrated several motors. By the close of the century steam engines would still be predominant, individual electric motors would very rarely be found incorporated into the design of machine tools. In 1899 five percent of industrial prime movers were electric motors, by 1919 eighty-one percent. Progress was slow largely due to the patent restraints. Another problem for machine tools was the early motors were large and lacked the necessary power that was required for heavy cuts at low speed.

The Anglo-French team of Lucien Gaudard and J. D. Gibbs demonstrated their alternating current transformer. Five years later the rights would be acquired by George Westinghouse and the Westinghouse Electrical Co. that he founded in 1886.

In 1882 Sir Charles William Siemens In his presidential address to the British Association proposed the watt as the unit of power.

The Yugoslav born American physicist and electrical engineer, Nikola Tesla, invented the first alternating-current motor. In May of 1888 he was issued several key patents for ac motors and related equipment which he sold to the Westinghouse Electric Company for $5,000 in cash. The contract gave Tesla $2.50 for every horsepower of electricity sold. Twenty years later most gear cutting machinery was still belt driven. When in 1886 Tesla patented a polyphase system of using alternating current to drive motors, which was ideal for heavy industry. the motor spindle was connected by a gear train of spur, helical and worm gears.

Welding: The Russian N.V. Bernardos patented carbon arc welding in 1877 which would come into its own when coated electrodes were invented in the early 1920's. In 1886 the English-born American inventor, Professor Elihu Thomson, was issued U.S. Pat. No 347140-42 for an electrical resistance welding system. He co-operated in 700 patented electrical inventions which included three-phase a.c. generators. The oxy-acetylene torch was invented by Edmund Fouche in 1900 at about the same time Johann Goldschmidt developed acetylene/oxygen welding in Germany.

At the turn of the century portable motor drills, rock drills, hoists, brakes, vehicles, revolving cranes were in use all powered by electricity.

Gas Engines: The Great U.S. Exhibition 1843 included one gas-engine by Drake. A French engineer and inventor, Jean Joseph Etienne Lenoir, patented in France the first successful gas-engine in 1859. The Lenoir Engine could work continuously under industrial conditions. The engines were used to drive machine tools. Lenoir also drove his stationery engine powered vehicle ten miles using street gas as fuel in 1863. The engine was preceded by the joint effort of the Italians Barsanti and Matteucci.

In 1875 a four horsepower vehicle was driven in Vienna by the designer Siegfried Marcus. Also, in 1875, Nikolaus A. Otto built "The Silent Otto", his horizontal gas-engine based on the four stroke cycle now known as the Otto Cycle. He used the four stroke principles outlined by the Frenchman, Alphonse Beau de Rochas, in his published work of 1862. In the next seventeen years 50,000 engines would be sold by Otto and Langen. Two of their assistants were Daimler and Benz.

The acknowledged pioneer of gasoline-engines was German engineer Gottlieb Daimler, who had worked with Karl Friedrich Benz of Mannheim. Daimler, patented his first gasoline engine, a single cylinder air cooled vertical engine that operated on the Otto principle in 1885. The Benz Patent Motor Co. received patent DRP 37435 for a "Vehicle with gas engine operation", January 29th, 1886, and it is regarded as the automobile's birth certificate. Karl Benz designed his three-wheeled vehicle with a four-stroke engine and electric ignition. Four years later Gottlieb Daimler attached his engine to a four-wheeled coach, the first "horseless carriage". The Benz car of 1886 is sometimes credited with being the first motor car. Benz installed a gearbox in 1899. M. Levassor and René Panhard obtained the French patent rights for the Daimler' engine and adapted it for highway use. This four stroke internal-combustion engine vehicle, like most automobiles today, had electric ignition, water cooling and a differential gear. "The drive was taken through a clutch to a set of reductions gears and thence to a differential gear on a countershaft from which the road wheels were driven by chains."

At the Philadelphia Centennial in 1876 a Bostonian George Brayton exhibited a gasoline engine vehicle along with six Otto gas engines. Brayton's engine had superior features and impressed a lawyer from Rochester named George B. Selden. In 1879 a U.S. patent would be granted to Selden which claimed to cover all gasoline powered vehicles. Most of the early manufacturers paid Selden a royalty, only Henry Ford refused to pay, and Selden lost the ensuing law suit in 1911.

In 1892 the diesel engine was patented by Christian Karl Rudolph Diesel, in Britain, and the first engine was successfully manufactured in 1897. The Paris born, but German engineer, Diesel, demonstrated the first practical compression-ignition engine. His diesel engine had twice the efficiency of comparable steam engines. Diesel disappeared from the Antwerp-Harwich mail steamer in 1913.

Materials: Great advances would be made during the 19th century in understanding the use, manufacture and heat treatment of metals. Many new materials would come into use and by the middle of the century wood would no longer be the primary gear material. The early part of the century would see a gradual replacement of wood by various metals, for example iron hulls for ships would appear in mid-century. Sir William Fairbairn's London plant built 120 iron ships in fourteen years commencing in 1830, and Sir William also published an important work "Mills and Millwork" in 1863 in which are described his replacement of all cast-iron shafts with wrought-iron.

In 1802 British patent No. 2707 was issued for Samuel Lucas's invention that made iron castings malleable by prolonged baking on a bed of metallic oxide. Another notable development of the 19th century was the British patent No. 5701 issued in 1828 to the Glaswegian J. Beaumont Neilson for introducing the hot air blast to forges and furnaces. This increased the reduction of refractory ores so that three to four times the quantity of iron could be produced with one third the fuel.

In Germany for the first time, the Geissenhainer process, a method of using anthracite coal to smelt iron was patented (1833). This process would have a major effect on U.S. production where anthracite was plentiful.

Circa 1820 Karsten established that the difference between pig iron, wrought iron and steel depended upon their carbon content, and it was only in 1831 that the German Professor of Chemistry, Baron Justus F. von Liebig developed a precise method for determining the quantity of carbon in steel. Volume 1 Holtzapffel's 1843 book is devoted to the subject of materials. On the determination of carbon content he wrote: "For the mode of analysis for ascertaining the quantity of carbon in cast-iron and steel, invented by M.V. Regnault, Mining Engineer, see Annales de Chimie et de Physique, for January 1839: also Journal of the Franklin Institute, vol. 25, p.327. It is stated that the analysis is very easy and exact, and may be completed in half an hour." The carbon contents of metals were listed by Holtzapffel with reference to Mushet's Papers, p.526:

Iron sem-steelified	contains 150th carbon
Soft cast-steel capable of welding	" 120th
Cast-steel for common purposes	" 100th
Cast-steel requiring more hardness	" 90th
Steel capable of a few blows	" 50th
First approach to a steely granulated fracture	" 30th to 40th
White cast-iron	" 25th
Mottled cat-iron	" 20th
Carbonated cast-iron	" 15th
Super-carbonated crude iron	" 12th

The beginning of the "Steel Age" is said to have been marked by Sir Henry Bessemer's presentation of his paper" The Manufacture of Malleable Iron and Steel without Fuel" at the British Association meeting in Cheltenham, August 1856. Prior to 1856 there were only two ferrous metals in general use, cast and wrought iron, a third carbon steel was in minimal production, Bessemer added a fourth, mild steel, to be followed in 1868 by Mushet's alloy steel. Steel was produced by the open-hearth process (Siemens-Martin process) using a regenerative furnace devised by Sir Charles William (Karl Wilhelm) Siemens and patented in 1856. He was a German born naturalized British subject. In 1861 the design was further developed in France by Pierre Émile Martin, the French metallurgist. The open-hearth furnace became the most widely used in the world.

In 1855 the Bessemer process that converted pig iron into steel at low cost had been patented in England by Sir Henry Bessemer. The process was only successful after using Robert Forester Mushet's British patent No. 2,219 of the same year. Mushet added ferro-manganese to the blown steel. Problems arose with the quality, leading to the open hearth process in 1861. A popular vote in "Scientific American" reported on July 25th, 1898 that the Bessemer process provided the greatest benefit to mankind of all inventions of the past fifty years. William Kelly, aided by four Chinese steel experts, in Eddyville, Kentucky, had also subjected molten metal to a blast of air in a specially constructed furnace. A lack of funds prevented him from obtaining U.S. patent (No. 17,628) until 1857 which was ten years after his furnace was built.

The English metallurgist, Sir William Chandler Roberts-Austen (1843-1902), published the first iron-carbon phase diagram in 1874. The iron carbide Austenite was named for him. Following the publication of Sir Roberts-Austen first iron carbon phase diagram he switched to martensite for the hard quenched phase and austenite for the high-temperature solid solution. The French metallurgist Floris Osmond used the term martensite for the high-temperature phase. The English metallurgist Sir Robert Abbot Hadfield found that by adding 12 -14 percent manganese and subsequent heating and quenching it would produce an alloy that was strong, hard and suitable for machining. Hadfield patented the steel in 1883.

In 1879 Sidney Thomas, a mere London police court clerk presented to the British Iron and Steel Institute a practical method for removing phosphorus from steel. This technical break through was so important the Bessemer Process became known as the Thomas Process. Carnegie who was present when the method was introduced purchased the rights to manufacture in the U.S. The method allowed the use of high phosphorus ore fields, anthracite and coke could now replace charcoal in the blast furnaces.

The "Harvey Process" covering the cementation and case-hardening processes were covered by U.S patents 376,194 in January 1888 and patent 460,262 in September 1891, other important U.S. patents of the century covered chromium, manganese and nickel steels. Isaac Babbit the American goldsmith also patented Babbit Metal in the U.S. in 1839 and Britain in 1843.

The first great American metallurgist Howe rhetorically asked "What is steel?" in 1875, no one had a definition but all had an opinion. Henry Marion Howe published his book titled "Metallurgy of Steel" in 1891. Howe, then president of A.I.M.E., proposed in a Chicago meeting the names ferrite, cementite, and pearlite. In 1895 he was to coin the name martensite, and in a 1903 letter to Sauveur the word eutectoid. Martensite was named for Adolph Martens (1850-1914). Albert Sauveur, circa 1890, was the only internationally recognized U.S. metallurgist in the steel world. He collated information on the behavior of steel from all over the world, analyzed the data, experimented and it was said "created order out of chaos"

A U.S. patent No. 211 938 was issued in February to a John Shaw for annealing and tempering with electrical heating. Previous to 1880 date a variety of quenching media such as whale oil, fish and animal oils had been used. E.F. Houghton introduced the first petroleum-based quenching oils and 125 years later the company he started, Houghton International, Valley Forge, PA., is still a leading authority on quenchants.

The first cold forging of steel by cold heading was used to convert wire into nails, followed in 1874 by bolts and screws. The extrusion of steel was regarded as impossible until the nineteen thirties.

Bronze: In 1854, in France, there was a major breakthrough in the improvement of bronze. The alloy of tin and phosphorus are similar to what is in use today. In May 1871, U.S. patent #115,220 was issued to Levi and Kunzel, and in August U.S. patent #118,372 to Lavroff, both were for phosphor bronze material. Roberts-Austen and Stansfield determined the phase diagram of the copper-tin bronzes in 1895.

Plastics: English chemist, Alexander Parkes, developed and patented the first man made plastic, xylonite in 1855. At the 1860 World Exhibition in London Parkes introduced another new material he named parkeskin. This was the first plastic material offered to the public, unfortunately, it was impractical outside the laboratory. The modern history of plastics can be said to have begun circa 1860. A substitute for ivory was sought and work began on natural resins. The first success with natural resins occurred in 1895 with a shellac material for recordings.

Tool Steels: We have previously mentioned the work of Huntsman in the last century and his development of crucible steel resulting in carbon steel cutting tools. In the early part of the century the only kind of tool steels had carbon contents between 0.85 and 1.5 percent. Faraday experimented with alloy steels in 1819 at the Royal Institution laboratory. He used a small special design crucible furnace, fired with coke and using hand held bellows he achieved high temperatures. Analyses a hundred years later showed he added copper, chromium, nickel, silver, gold, platinum and rhodium. Not enough steel was produced to run mechanical tests. His alloys were used for cutting tools on a small scale but being ahead of their time never gained wide usage. Faraday is said to have developed nickel steel but the credit is given to Johann Conrad Fischer. He first produced the steel at his works in Schaffhausen, Switzerland, in 1824-5. in 1839. J.M. Heath's invention of adding manganese to make a better tool steel was very successful. The manganese improved the quality of Huntsman's crucible steel. Heath had to industrially produce the manganese from a mixture of manganese oxide and coal Faraday had shown in 1819 that "pure manganese could be obtained by attacking its oxide with a mixture of oil and charcoal".

Tungsten steel was invented by Josef Jacob and Franz Köller in 1855 and made at Reichraming, Austria. Major improvements in tooling were made with the birth of tungsten-high-strength steel. Application tests proved the possibilities and "Wolfram" steel for boring and turning tools became popular.

In the mean time, R. F. Mushet, the English metallurgist, worked in secrecy on improving the Bessemer process that produced brittle steel. His father David Mushet had earlier made the Bessemer process of low cost steel practical by adding spiegeleisen, a Prussian iron ore rich in manganese and directly fusing it with malleable iron and charcoal. The method was only applicable for special limited use. A patent for the process was issued to David Mushet in 1800 for the improved cast-steel. In 1868, at Coleford, Gloucestshire, R. Mushet discovered that by adding small amounts of tungsten high quality steel could be obtained. This steel was invented as "R. Mushet Special Steel" and did not require water quenching and, self hardened in air. Mushet's actual production method is still secret. This was the first steel worthy of the name tool steel. "There is no doubt that Mushet did more than any others to perfect the production of tungsten steel for tools."

The working life of a cutting tool is limited by its maximum cutting speed. In 1850 the carbon steels were limited to a maximum cutting speed of 40 fpm. The high-carbon tungsten-manganese and vanadium steel, was the forerunner of the whole family of tool steels The steel enabled cutting speeds to be increased to 60 fpm. Chrome steel arrived on the scene in 1877 and was commercially produced by Jacob Holtzer at Unieux, France. Frederick Winslow Taylor and Maunsel White the chief engineer at the Midvale Steel Company in Philadelphia, in close cooperation with Bethlehem Steel Works developed high speed steels over the three year period 1898 -1901. They added high portions of chrome and tungsten introducing the new high-speed steel at the 1900 World Exhibition in Paris. Visitors were amazed to see a tool with a red hot tip peeling away cuttings at blue heat. By the end of the century cutting speeds had trebled reaching 120 fpm and 150 fpm was common in 1912. Taylor and White steels are still in use today in practically every worldwide machine shop. Taylor thoroughly analyzed the work process and presented his findings in New York in 1906 at an ASME meeting under the title "The Art of Cutting Metal", which was reprinted in the "Iron Age". This most important paper was based on the results of 50,000 experiments in which 400 tons of metal had been machined. Started with the support of Sellers the experiments were to cost over $200,000 in a twenty-six year period. In the first year, using Mushet's steel, Taylor established that a round-nosed tool could run at a faster speed than the hitherto diamond-point. His next discovery was that a constant stream of water at the cutting point allowed for a thirty to forty percent increase in speed. This was in direct opposite to Mushet who said his self-hardening tools must run dry. To prevent corrosion carbonate of soda was added to the water and, when new machines were installed at Midvale, they were equipped with "suds" tanks. Although there was no attempt to keep this secret, it would not be copied elsewhere until 1899. Another important discovery was to disprove the fallacy that self-hardening tools should only be used for hard metals. Taylor proved that the greatest gain, ninety percent, was on cutting soft metals, whereas the gain in cutting time was only forty-five percent with hard metals. He also discovered that the feed power should equal to the drive power, and considered twelve variable factors are involved in metal cutting. Taylor found difficulty in expressing solutions by formula and in consequence built a slide-rule for production engineers providing accuracy in place of the previous rule of thumb. Their work continued and will be discussed in chapter six.

Testing Laboratories: Friedrich Wöhler, the German chemist who isolated aluminum in 1827 and discovered calcium carbide from which he obtained acetylene, advocated in Germany the setting up of state owned material testing laboratories. In England they would be set up by private enterprise. The first successful British laboratory was built by David Kirkaldy in Southwark, London in 1865. The first laboratory on the continent was founded in 1871 at the Polytechnical Institute of Munich, and the next by Martens at the Berlin Polytechnicum. Laboratories soon followed in Vienna, Zurich, Stuttgart, Stockholm and St. Petersburg.

During the 1889 International Exposition in Paris an International Congress of Applied Mechanics was also held. One section was devoted to the testing of metals. There was also interest in standardizing tests procedures and discussion on establishing an international society for the testing of metals which would lead to several International Congresses.

In Sheffield, England, Henry Clifton Sorby (1826-1908), geologist and metallurgist pioneered the microscopic examination of iron and steel by reflected light. He adapted his microscope technique from rocks to metals by treating polished surfaces with etching materials. He critiqued the newly published works of the leading metallurgists, Martens of Germany, and Floris Osmond from France. With the increasing use of iron and steel it was realized that only knowing the ultimate tensile strength was inadequate, and not a measure of the quality.

The introduction of "high-speed" tool steel and its use in machine shops resulted in many test programs. It was necessary to have precise knowledge of the wide variation in cutting properties and the forces acting upon the tool. Tests were performed on lathe tools, drills, cutters, files and hack-saw blades. Many different types of commercial testing machines would be designed between 1890 and 1925. A typical testing program took place in1898 and was printed in "Engineering Association Lehigh University Journal" by Professor l. P. Breckenridge's on the results of his investigations to determine the pressure on a drill's point when drilling medium cast iron.

Elasticity/Fatigue/Tensile/Poisson's Ratio: Dr. Thomas Young (1773-1839), considered possibly the most gifted and versatile of British scientists, introduced his concept of the Modulus that bears his name, stress divided by strain, which measured elasticity. Young's "Modulus of Elasticity" published in 1807 was the first definition of elasticity, "the measure of resistance of materials to deformation under stress". Young also introduced the physical concept of energy being also the first to use the word energy with its present meaning in physics. Included in the Design section of this chapter are his conclusions on gear pressure angles.

With the development of the steam engine fatigue failures of moving parts was a common occurrence. In England, W.A.J. Albert carried out the first experiments to determine the limits of fatigue failure in 1820. A.Z. Wöhler in 1860 published in Germany the results of his classical fatigue experiments. These were the first fatigue tests on different metals in which the magnitude of the applied loads was carefully controlled.

In 1829 the French professor of mechanics at the Sorbonne, Siméon Denis Poisson, had introduced "Poisson's Ratio", the constant of proportionality between primary and secondary stress. Professor Rankine introduced the term "Stress" for strain or latent force, and writers of the day urged the wide spread adoption of this term.

In 1868 E.R. Walker, Newcastle under Lyme, England, published safe working stresses for cast iron and steel that became known as the "English

Rule". Wilfred Lewis would use this data in his 1892 formula. A series of mechanical tests on iron and non-ferrous materials were performed for the U.S. Army in 1840-60. In 1872 the government appointed a U.S. Board for the Testing of Iron, Steel, and Other Metals. Their first report in 1880 contained the known mechanical properties with detailed stress/strain curves.

Tinius Olsen, an immigrant to the U.S. from Norway, patented the *"Little Giant"* in 1880. This was the first universal testing machine that in a single unit could evaluate a material's tensile, transverse, and compression. The machine provided a graphical record of the measurements. The fundamentals of the machine became the basis for all future testing machines produced in the U.S.A. His system of screws and gears could apply controlled forces to the test piece. An optional attachment used an electric motor to automatically move the weight on the beam. The *"Little Giant"* won gold medals at the 1881 Atlanta and Cincinnati industrial expositions. Tinius Olsen, Inc., Horsham, PA., is still an active company in the same field managed by his descendants, and providing innovative solutions to problems in testing.

Friction: The Scottish engineer George Rennie, superintendent of the mint undertook an impressive series of experiments on friction that concluded with a detailed report in 1829. In addition to testing eleven combinations of wood all available metals were also tested. He metal tests involved the effects of load, surface area, sliding velocity and lubricants.

In 1823 Arthur Jules Morin, a French Army captain, carried out friction experiments using heavily loaded blocks of different materials. He agreed with Coulomb's findings that resistance to rolling was directly proportional to load and inversely proportional to the radius of the roller. Morin made carefully conducted experiments with sliding friction at the Metz Engineering School, and reported the results in "Mèmoires de l' Institute and "Frottement des axes de rotation" in 1833. Further work was carried on by Thurston, Westinghouse, and the Research Committee of the Institution of Mechanical Engineers directed by Beauchamp Tower.

In the period 1885-1888 Tower carried through important friction tests for The Institution of Mechanical Engineers Research Committee on Friction at High Velocities. His accidental discovery of substantial pressures in the oil film is regarded as an important milestone in the history of tribology. Alex B.W. Kennedy in his 1886 "Mechanics of Machinery" thought it was misleading to use the much-used phrase "co-efficient of friction", preferring a friction-factor for the given pair of surfaces allowing for the ratio, and the frictional resistance of the surfaces to the pressure causing the friction.

Professor Robert H. Thurston, the first president of ASME, studied the Laws of Friction. He delivered a series of lectures to the Master Car Builders Association which were published in the Railroad Gazette. They were the basis for his 1855 book "A Treatise on Friction and Loss Working Machinery and Millwork." Gears are amongst the many subjects covered.

Efficiency: The idea of a machine's efficiency began with Moseley writing in the Philadelphia transactions of 1841, and "Mechanical Principles of Engineering" in 1843. He termed it the modulus of the machine and worked out the values for spur and bevel gearing.

Thurston stated "... the loss of power in mills, with different machines, from 5 to 90 percent..." The losses of fifty percent of the power in overcoming friction between lubricated surfaces in mills, three to sixteen percent in steam engines, and approximately fifteen percent in metal working tools forced research in reducing friction.

W. Lewis undertook the most extensive experiments to determine the efficiency of cast-iron spur, helical, and worm gears for the Wm. Sellers and Co. The worm gears tested had thread angles of 5, 7, and 10 degrees. The results were described in detail by Lewis in the Transactions of ASME, vol. vii. Thurston and Sellers both concluded that helical gears on parallel shafts were as efficient as spur gears. Grant also concluded that the difference in efficiency between cycloidal and involute teeth was minute. Grant was very critical of non-parallel gears and believed that they were only suitable for light loads, and that they "waste from a quarter to two-thirds of the power".

In Philadelphia, in the 4th edition of "The Constructor", gear efficiencies were analyzed by Professor Franz Reuleaux in 1893. In their 1894 book "The Mechanics of Machinery", the German authors Professors Weisbach and Herrmann also analyzed gear efficiencies. An important work on gear theory was also produced in Germany by Herrmann in a section of Weisbach's 1898 book "Mechanics of Engineering and Machinery". The book was translated by Professor Klein, Bethlehem, Pa.

Hardness: The first hardness scale was published in 1820, the work of the German mineralogist Friedrich Mohs. The comparative system is still used to provide an approximation of the hardness ranking. The method is based on a sequence of ten defined substances ranging from diamond to talcum and their relative ability to indent or scratch a surface. Hardened tool steel would range between a seven and eight on the scale.

In 1865 the Russian metallurgist Dimitri Tschernoff's work proved that carbon steels had to be heated above a specific temperature in order to be hardened by quenching. Maximum hardness was only achieved by cooling above a critical rate below 200° C. He used a file to measure relative hardness, and colors to describe high temperatures. Brinell in Sweden published similar results in 1885.

A report to the French Commission of Material Testing by Martel in 1893 introduced the first useful procedure for dynamic testing of hardness. In 1897 Foeppl advanced the work of Rèaumur by measuring the area of contact of two semi-cylindrical bars when pressed together and dividing this area into the load to evaluate their hardness.

Measurement: Prior to Whitworth the standard measurement practice was to use inside or outside calipers set to an accurately graduated rule. In 1851, at the Great Exhibition, Sir Joseph Whitworth exhibited a measuring machine that could measure to an accuracy of one millionth of an inch. The machine is illustrated in figure 5-16. In actual practice the device was for the personal use of Whitworth and he could compare dimensions to 0.0001 inch.

Figure 5-16 Whitworth Measuring Machine

This was a major improvement over 1776 when Wilkinson had bored a 50-in. cylinder that "doth not err the thickness of an old shilling." He advocated using end measure of length. Also in 1856 Whitworth introduced plug and ring gages. They were made as standard sets in various dimensions and were expensive and only useful for cylindrical work In the U.S. during the 1870's the cheaper and more practical "Go" and "No-Go" gages came into wide use.

Professor William A. Rogers of Harvard and George M. Bond were commissioned by the Pratt and Whitney Company to establish standards. They rejected Whitworth's principle of measurement and used a linear measurement with a microscope that had micrometer adjustment of the eyepiece. Their "Comparator" could check any gage against master standards and avoid the wear of measuring faces that was present with the previous systems.

By 1880 very few machine shops had any special machines for precise measurement. Smaller engineering projects could be measured by hand tools similar to what are shown in figure 5-18 and even special set-ups as shown in figure 5-17.

Figure 5-17 Late 19th Century Bench Micrometer
Built by John Holroyd Co.

In France in 1848, J.R. Palmer produced a micrometer similar to what we use today. In 1850 Joseph R. Brown of the Brown and Sharpe Company, invented the vernier caliper, it was the first practical tool for exact measurements which could be sold in any country at a price within the reach of an ordinary machinist. (The scale used was invented by Vernier in 1631). J. R. Brown and Lucian Sharpe saw a Palmer gage at the Paris Exhibition in 1867 and recognized its importance. In 1869 they had a small unit accurate to 0.004 on the market for measuring sheet steel. In their 1871 catalog Brown and Sharpe used the term micrometer for the first time. By 1877 they had the one inch micrometer available and within three years they were in practically every machine shop in the U.S.

Figure 5-18 Shop Measuring Instruments circa 1880

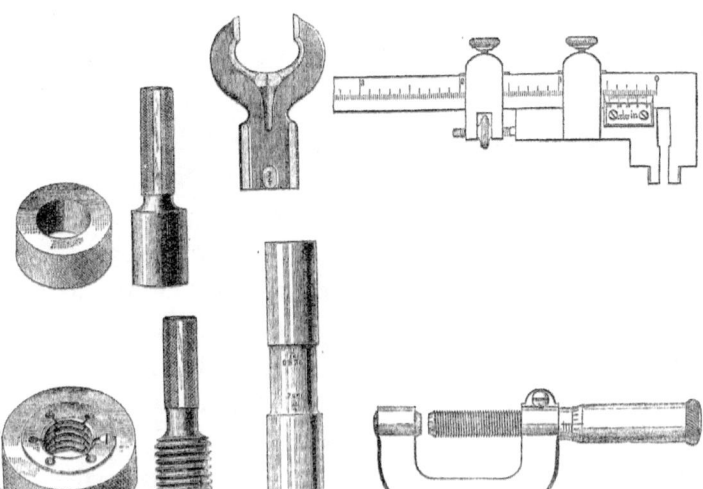

To produce accurate gears Brown and Sharpe used a copper ring that was graduated at the office of the Coast Survey in Washington. The graduating machine had 4340 graduations that had been copied from the plate of the Troughton and Simms machine in London, England.

The next important introduction to improve measurement were gage blocks, credited to the work of the Swedish machinist C.E. Johansson.

In 1880 William Edward Ayton and John Perry in England invented a machine for measuring horse-power that later became known as a dynamometer. Their machine remained in popular use for the next fifty years. The true ancestor of the dynamometer is considered to be Sellers dynamometer of 1907.

Metric System: In 1800 the English astronomer Sir John Frederick William Herschel split sunlight with a glass prism, installed thermometers and found that energy increased towards the red band. He then discovered that the highest temperature was just beyond the red band, and had a supposition on the existences of infrared light. It would only be in 1931 that this knowledge was put to use in radiation sensors. During the early 1940's, this knowledge resulted in the first present day infrared quantum sensors that are now essential tools in the production and inspection of metals.

The French physicist Jacques Babinet standardized light measurement by using the red cadmium line's wavelength, and the standard for the angstrom unit. This enabled an accurate measurement of the meter in 1829. The primary standard, a platinum-iridium bar with three fine lines at each end, the distance between the middle lines at each end at a temperature 0° C is by definition one meter, and is kept at the International Bureau of Weights and Measures near Paris, France. Two copies are retained at the Bureau of Standards, Washington D.C. The Metric Convention formed an International Bureau of Weights and Measures in 1875.

In 1837 the official metric system was established in U.S. by the Law of the Meter. In the U.S. imperial units, yard, foot and inch were used, but their definition was in terms of the meter. Until 1960 the yard was defined as 36/39.37 meter. Due to the ratio chosen the U.S. yard was longer than the Imperial yard by some 5 millionth of an inch. The legal equivalent for commercial purposes was fixed at 39.37 inches by law in July, 1866. In 1877 Brown and Sharpe had their standards checked with the Washington standard: "Taking 39.370 as the standard, there is only 0.00023 in. in the meter difference to our comparison…" By executive order in 1893 it was authorized to derive the yard from the meter.

Calculators: Viewed as the pioneer of the modern computer, Charles Babbage, *"difference engine"* was intended to calculate logarithm tables. He was awarded the Astronomical Society's gold medal in 1822 for his achievement. The English mathematician designed the first practical calculating machine. By using a series of gear wheels repeated additions were performed and printed to thirty-one significant figures. Joseph Clement built the machine and because of the accurate workmanship it would become his most renowned work. Babbage's design for a punch card programmed machine could not be built with the available machinery.

In Washington in 1889 the statistician Herman Hollerith was able to compile and record data by punching holes in cards that were then run through electrical machines. Hollerith formed a company to manufacture these machines that in 1911 became a part of the Computing –Tabulating - Recording Company, later known as the International Business Machines Corporation.

Machine Tools: During the early part of the 19th century mechanical equipment was built almost entirely by hand. The gradual introduction of machine tools accelerated the manufacture of metal components. Machine tools would be considered "the industry of industry." By 1830 all the principle machine tools with the exception of the surface grinder and the broach were in a recognizable form. To this time most of the innovations were the result of those who had worked with Maudslay. Between 1830 and 1850 Britain would be fully occupied with the improvement of machine tools, to be followed by intense industrialization. Machine shops were systematically setup with the machine-tools in neat rows. By 1850 the general principles of machine tools as we know them today had been established. To this date machine tool building had been almost exclusively British. In 1853 Sir Joseph Whitworth joined a Royal Commission visiting the New York Exhibition and reported that American machine tools were generally inferior to the British, but their eagerness to use machinery whenever possible in place of man-power appealed to him. Whitworth of Stockport, England, was the most famous and successful of all the engineers who worked with Maudslay. He replaced Maudslay as the leading figure in the history of machine tools, and the firm he founded became the world's leading machine tool manufacturer. Whitworth was a toolmaker who in 1835 set up a plant with the intention of making precision machine tools for sale. Sir Frederick Henry Royce of Rolls-Royce fame became known for craftsmanship so Whitworth's name guaranteed fine workmanship, performance and precision. He was also a strong advocate for standards and decimalization. Standardization and mass production were not possible without measurement standards. He introduced the box design for machine-tools with hollow-frame construction thereby increasing the stability and rigidity. By 1850 his lathes, planers, drillers, slotters, shaping and gear cutters were considered to be the standard for machine tools all over the world. Previously tools had been made by the user for their own specific purpose. In the Exhibitions of 1851 and 1862 Whitworth had twenty-three exhibits. He sought to standardize screw threads and his proposals became the British thread standard in the 1860's. The standard was responsible for improved machine tools and gear cutters.

Sir Fairbairn, author and major ship builder, neatly summarized the revolution in machine tools over the previous fifty or so years. In his Presidential address to the British Association, Manchester, 1861, he stated: "When I first entered this city the whole of the machinery was executed by hand. There were neither planning, slotting nor shaping machines; and with the exception of very imperfect lathes and a few drills, the preparatory operations of construction were affected entirely by the hands of the workmen. Now, everything is done by machine tools with a degree of accuracy which the unaided hand could never accomplish. The automation or self-acting machine tool has within itself almost creative power; in fact, so great are its powers of adaptation that there is no operation of the human hand that it does not imitate."

Fairbairn realized that the having a low speed main drive and increasing the peed to the machine by countershafts was inefficient. He wrote essays noting that for the transmission of a given power the weight and strength of shafts and wheels can be reduced in direct proportion to the increase in speed. In 1818 his principles were used for the first time in a new mill owned by MacConnel and Kennedy. The system was quickly adopted and lasted until the electric motor made line-shafting obsolete.

Samuel Smiles wrote: "In the course of a few years an entire revolution was effected in the gearing... the speed was increased from 40 to upwards of 300 revolutions a minute". This speed increase made it necessary to improve the design and manufacture of gears.

The French engineer and geometrician, Jean Victor Poncelet, wrote a detailed report on the machinery and tools exhibited at the 1851 International Exhibition. London. He wrote a treatise on practical mechanics and a memoir on water-mills in 1826. As professor of mechanics in Metz he had an influence on gear design.

The Government of Napoleon 111, concerned about the importation of British machinery, developed a program for industrial expansion. The successful results were easily discernible by the machinery exhibited in the 1855 and 1867 Paris Expositions.

Johann Georg Bodmer was born in Zurich and is considered one of the foremost inventors and mechanical engineering geniuses of the nineteenth century. His list of patents took up over eight pages and ranged from machine tools to an opposed-piston steam engine. Bodmer's British patent #8070 in 1839 "Tools, or Apparatus for Cutting, Planing, Drilling and Rolling Metal," filled fifty-six pages of what was virtually a mechanical catalog. He designed and equipped a machine tool factory in England with all new equipment and traveling cranes, which by its layout was designed to save labor, energy, and movement. The later principles of Henry Ford "to place the tools and the man in sequence of operations," was already in place at Bodmer's plant. Between 1830 and 1850 his main work would be in the improvement of machine tools.

At the midpoint of the century the dominance of the English machine tool makers ceased, being overtaken by the New England tool builders. The growth of cotton and the law of 1785 were two of the main factors for the growth of New England machine builders. During the next seventy-five years America was to produce the basic modern machine tool, including the universal miller and grinder. Originally all tool builders in the U, S. were east of the Allegheny but between 1850 and 1900 Cincinnati became the machine tool capital of the world. At the industry's peak some 15,000 were employed in forty machine tool companies. By 1915 Cincinnati built machine tools accounted for thirteen percent of all machine tools that were built throughout the world.

In Cincinnati five men and the companies they founded were considered the leaders in the quality and design of machine tools. An English immigrant John Steptoe was probably the first to build machine tools in the city. George A. Gray (1839-1905) a designer, formed the G.A. Gray Company to make lathes, and had previously started the Universal Radial Drill Company. Richard K. Leblond founded the Leblond Company in 1887. Another English immigrant William Lodge with William Davis as partner started Lodge and Davis in 1891. The Cincinnati Screw and Tap Co. begun by Fred Holz and George Mueller in the 1870.s and the company name changed to Cincinnati Milling Machine Company in 1889. Frederick A. Geier had joined the company in 1887 and his skills assisted in making them the world's largest machine tool manufacturer. The National Association of Manufacturers (NAM) was formed in Cincinnati in 1895.

Joshua A. Rose book, "The Complete Practical Machinist", was published in Philadelphia in 1875. The work was so popular that by 1894 nineteen editions had been printed. Information on cutting tools, heat treatment, tool grinding, grinding wheels, and the currently used machine tools were included. All the machine tools illustrated were driven from a line shaft or by foot pedal.

Johann Zimmerman, Chemnitz, Germany was the first German manufacturer to specialize in machine-tool manufacture. At the 1862 World Exhibition he exhibited twenty machines.

Lathes: In the 1840's when Britain was producing heavy production lathes most American lathes still had a wooden bed with iron-plated guide ways.

Figure 5-19 Foot-Powered Lathe

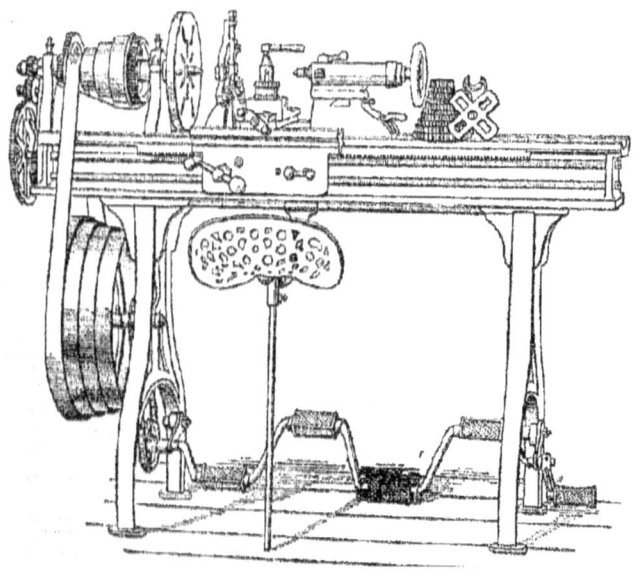

The simplicity of corresponding American designs of the period can be seen by comparing figures 5-19 and 5-21. The foot powered lathe was built by W.F. and John Barnes Company, Rockford, Illinois. The lathe in figure 5-21 was built by Whitworth in 1839. Not all American lathes were as crude as can be seen in the pioneering work of Clements and Blanchard. Joseph Clement (1779-1844) came to London and worked for Bramah as superintendent of his works. He took a very important part in the development of lathes, planers and established a thread standard. Clement's facing lathe of 1827 is a major advancement in machine tools. Variable speed was obtained by using cross belts on cone pulleys.

Figure 5-20 Clement's 1827 lathe.

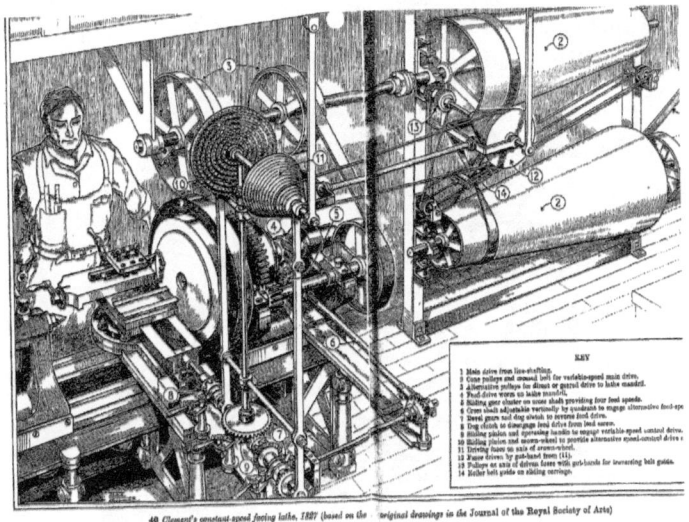

Figure 5-21 Whitworth Production Lathe 1839

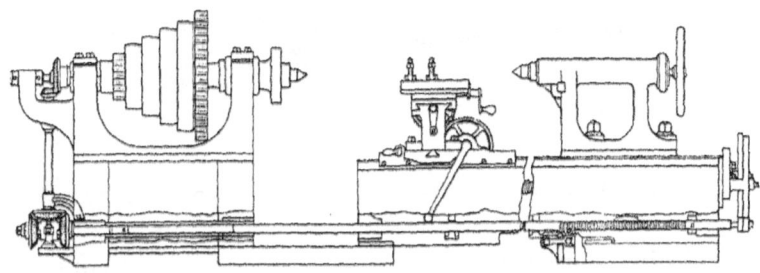

It would be in special designs that American ingenuity would come to the fore. One of the first of these innovative designs would be Thomas Blanchard's patented lathe for turning irregular forms that was first made in 1818. Blanchard's lathe patent expired in 1823. The lathe shown in Figure 5-22 was installed at the Springfield Armory and used for over fifty years. This was the earliest example of tracer control technology. Blanchard's lathe could automatically turn irregular forms which were an exact facsimile of the pattern. This was not the first machine that could turn irregular forms but it was the most useful up to that time. The lathe was also more mechanical than previous designs and would still be in general use in 1915. His lathe was more precise than any other of the metal-working machines in the armories. Several hundred Blanchard lathes were built by the Ames Manufacturing Company, Chicopee, Mass. The basic lathe design would lead to the design of a new range of machine tools such as milling machines

Figure 5-22 Blanchard's Lathe

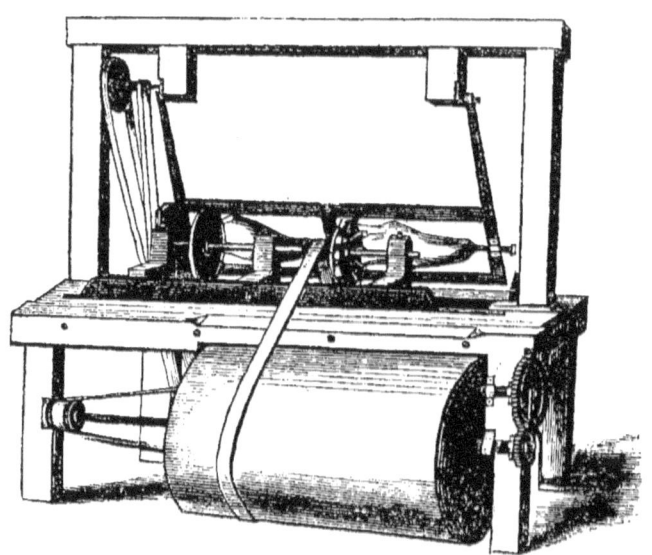

BLANCHARD'S LATHE.

Murdock's replacement at the Boulton and Watt Soho plant, William Buckle, built the first large screw cutting lathe, previously screws had been produced by hand. Important lathe developments had begun in the USA in 1840's. The turret lathe was invented in the U.S. by Stephen Fitch in 1845. Eight successive operations could be performed without stopping the machine to change tools. This was the most important advance in lathe design since Maudslay. Twenty five years later Christopher Spencer made it automatic and capable of performing operations untended. E.G. Parkhurst the renowned American machine tool engineer and inventor of the collett chuck wrote in 1900: "One only does justice to the memory of Stephen Fitch when one recognizes in him the creator of one of the most time-saving machines ever invented." Robbins and Lawrence Company, Vermont, built a fully developed an octagonal turret lathe upon which eight machining operations could be performed. The turret lathes were built commercially by the company in 1854. A British patent for a capstan-lathe with a vertical turret was issued prior to 1840. However, the turret lathe was not to be found in Britain, or even listed in the London World Exhibition catalog of 1862. Further capstan lathe developments in the U.S. in the decade up to '65, such as a ratchet-and-pawl mechanism to automatically rotate the turret are attributed to British immigrants F.W. Howe. He started his career as a mechanic in the Gay and Silver shop, moved to Robbins and Lawrence as a draftsman, and then rose to superintendent. When the company failed due to a rail car and arms contract he moved as superintendent to the Providence Tool Company. When at this company he designed a turret screw machine. The chuck could be revolved, indexed and swiveled. Parkhurst when reviewing his years of experience in the industry was of the opinion that no one had contributed more to the development of the metal-cutting lathe than Howe.

Figure 5-23 Hendy-Norton Gear Box

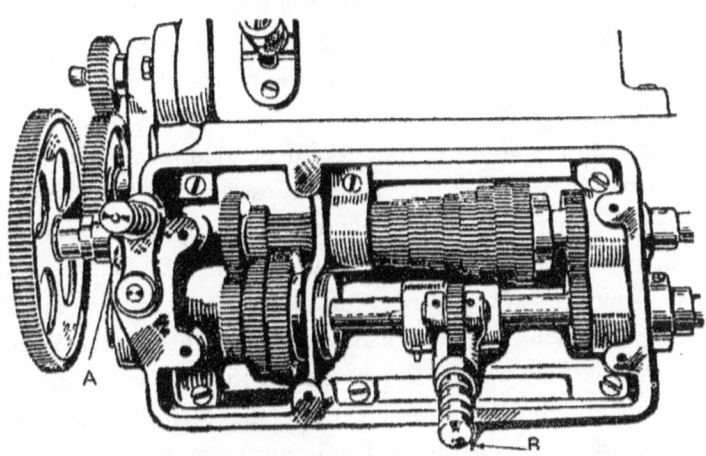

The method of changing speed by moving a flat belt to a different diameter pulley was improved with the introduction in the U.S. of a hand-change-gear box in 1892. The Hendey Machine Company of Torrington, Conn., built a high-grade lathe with a change gear box that would be applied to all types of machine tools. This was one of the most important contributions to machine tool design at that period of time. The first units were quickly improved and by 1900 a friction-drive had been incorporated. This modification allowed for the changing of the spindle speed rapidly and easily.

Pratt and Whitney monopolized the market for tool-room lathes until the Hendey top grade machine tool-room lathe with its quick-speed change box. A tumbler gear drive was invented in the U.S. by W. P. Norton in 1889. The revolutionary gear box such as the first Hendy-Norton also required many gears. Machine tools were themselves major gear users, and the designs made regular use of large worm gears in the heavy lathes, planers and borers. A gear-box for multi-spindle lathes was introduced in America by Taylor and White in 1892. Speeds were also selected with the movement of a hand lever.

In 1880 the 10th U.S. Census included a report on "Power and Machinery" in which F.R. Hutton reported on "Machine Tools and Wood Working Machinery". Hutton was the assistant in mechanical engineering at the Columbia College School of Mines, N.Y. On discussing machine tool drives he wrote "The worm-system is objected to by many builders on account of the danger to it if allowed to get dry by neglect...The wheel is made of cast iron, the worm of wrought iron, steel, or cast iron. In some cases it turns in an oil-pan...Very often....the feed screw is splined to carry the worm on the bevel- gear. This is objected to by some on the ground of the wear on the top of the thread and at the points of the threads where they are cut by the spline."

Planers: Every type of machine tool depended for its precision on a truly flat surface. The planer was therefore, essential to the advancement in machine tool construction, and only second in importance to the lathe. James Bramah, inventor of the hydraulic press in 1795 and among the first to propose using the screw propeller for marine drives, was an influential tool maker. Bramah built a planer using a cutter in the same manner as on a face-milling machine which was patented in 1802. Twenty-eight cutters were fixed in a horizontal disc which was rotated on a vertical spindle. In his Pimlico works he trained many future inventive mechanics including Maudslay and Clement. Claims to being first with the idea have been made by Matthew Murray and James Fox in 1814, and Richard Roberts in 1817. In America the absence of a planing machine meant the work was done by a cold chisel and hammer at least until 1839. Although planers were being built by Gay, Silver and Company as early as 1831. Fox made use of both rack and lead screw traverses. A freely sliding worm gear on the carriage was connected by spur gearing to the rack. The power driven table was automatically reversed by a combination of three bevel gears and a double-faced dog clutch sliding on a keyway between two gears. A planer in the Birmingham Science Museum, England, with an 1817 date is believed to have been built by Fox. He made improvements in the carriage traverse, and by 1817 had a powered table with automatic reverse and feed with horizontal and vertical tool movements making extensive use of bevel gears. His machines were so well received that they were sold and exported to France, Germany, Poland, Russia and Mauritius. Richard Robert's 1817 planer is in the London Science Museum and is very crude in comparison with Fox's machine.

In 1820 Clement was operating a planing machine with the work mounted on a reciprocating table to machine lathe guide ways. We have already mentioned his brilliant lathe of 1827 that would not be surpassed for sixty years. His second machine built in 1825 could plane a six foot square and would remain the largest in the world for the next decade. Another major difference was that Clement used two tools rather than the single tool of other tool builders.

The Scottish engineer James Nasmyth who held an 1842 British patent #9382 on the steam hammer, later transferred as a U.S. patent to S.V. Merrick in Philadelphia, held many other patents such as a worm-geared tilting pouring ladle, and a planing machine. The hammer was completely built at his plant in Manchester, England, and. made heavy forgings practical. In addition to various machine designs he made a significant contribution with his 1848 book "Remarks on Tools and Machinery". Until Maudslay's death in 1837 he was his personal assistant. His planers moved the table with a rack motion. Nasmyth realized that the planer with its reciprocating table was slow and cumbersome when used for other than large jobs. He built a small shaping machine that became his number one seller. The shaper, a modification of the planer, was popularized by Nasmyth circa 1835. He transferred the motion from the table to the tool. Similar in action to a horizontal steam engine it was nicknamed "Nasmyth's Steam Arm". He built a larger machine using a rack motion with the pinion having a limited amount of lateral motion. This lateral motion permitted the pinion to alternate with the outside and inside of the pinion's teeth, thus reversing the motion. Whitworth improved Nasmyth's shaper by adding the quick-return motion that became known as the "Whitworth Quick-Return Motion". Nasmyth made several innovations to shop practice and built his "double' or ambidexter" lathes that in 1837 introduced his mechanism that used two identical meshing idler gears mounted on a lever with a rocking pivot. The lever movement provided either a three or four gear driving train that allowed neutral or reverse position of the lead screw. As a perfectionist he was disdainful of the workmen and designed accordingly. He said his machine tools "never got drunk; their hands never shook from excess; they were never absent from work; they did not strike for wages; they were unfailing in their accuracy and regularity..."

Whitworth's first planing machine was designed to plane in both directions of travel with a single tool named "Jim Crow". The tool was automatically rotated 180° at the end of each travel. He later reverted to cutting in one direction only with an innovated quick return motion. The system used two belts with a shifter so a belt was alternatively driving or loose. The system was superior to the bevel gear arrangement used by Fox and others.

The Welsh mechanical engineer, Richard Roberts, established a machine tool business in Manchester in 1817 having previously worked for Maudslay and Wilkinson. Roberts improved on Maudslay's screw-cutting lathe, and built one of the first metal-planing machines (now in the Kensington Museum), a punching and shearing machine, and also invented the slotter and a key-seater. His planing machine and large industrial lathe incorporated a back-geared headstock to effect changes in spindle speeds. The crown gear device provided a choice in relative speeds between spindle and screw, forward and reverse and automatic stop. He built numerous machine tools including gear cutters. In 1833 Roberts was the first to use differential gearing on a powered road vehicle, a steam powered carriage. Roberts also built twin screw steamships. He was considered to be the most productive engineer of his time.

Nasmyth led a British committee on a tour of the New England plants in 1851. Contracts were placed with Robbins and Lawrence, Windsor, Vermont, for 150 machine tools, numerous jigs and fixtures, and with the Ames Company, Chicopee Falls, Mass., to completely equip an English plant based on the American System.

When the Philadelphia Mint was being retooled the horizontal steam engine was to be built and designed by Franklin Peale. He was assayer, melter and refiner in 1836, and chief coiner in 1839. With a Dr. Patterson they visited the Cardington Locomotive Works, Philadelphia, Pennsylvania, to see the operation of their planer. There were only two others in operation in the U.S., one at Kemble's West Point Works and the other at Dr. Nott's Novelty Works, Schenectady, N.Y. both imports.

The cold chisel and file were the main tools in providing a flat plate. The sixteen-ton bed-plate required in 1839 for the engines of the U.S.S. Mississippi was finished by this hand method.

William Sellers was born in Pennsylvania (1824-1905) and in 1848 opened a small tool-making shop in Philadelphia. Probably no one in America had a greater influence on machine tools than Sellers. He has been called the Whitworth of America, and was granted over ninety U.S. patents. Whitworth is reported to have said that Sellers was the greatest mechanical engineer in the world. In 1915 Joseph W, Roe in his book "English and American Tool Builders" wrote: "Almost, from the first Sellers cut loose from the accepted designs of the day. He was amongst the first to realise that red paint, beads, and mouldings, and architectural embelishments were false in machine design. He introduced the 'machine gray' paint which has become universal; made the form of the machine follow the function to be performed, and freed it from all pockets and beading. Like Bement* (a Philadelphia tool-maker) he realised that American tools then being built were too light and they put more metal into their machines than was the practice elsewhere. From the first he adopted standards and adhered to them so closely that repair parts can be supplied today for machines that were made fifty years ago." In the Vienna Exhibition, 1873, Seller and Company's lathes won first prize for quality and design. His best known machine patented in 1867 was a helical geared planer. Sellers adopted a tooth form using a twenty degree pressure angle, with an addendum of 0.3 and a clearance of 0.05 of the circular pitch. In 1866 Sellers produced an improved automatic miller for gears. Sellers also built planers, large milling machines with worm and rack driven tables, and gear cutting machines. The rack teeth were cut straight across and the worm set at an angle. The method was adopted by several manufacturers of large milling machines. Supported by thrust bearings, the worm ran in an oil bath. As President of Midvale Steel Company he encouraged and financed Frederick Taylor with his cutting tool experiments. Sellers presented his system of screw threads to the Franklin Institute in Philadelphia in 1864. He simplified Whitworth's design by using a thread profile of 60° in place of the more difficult to produce 55°. The previous rounded profile in the root was changed to a flat profile. Whitworth's thread was too coarse having been designed for use with iron. Sellers introduced a systematic approach to thread pitches, threads per inch, and depth which resulted in the U.S. National Thread Standard in 1868 and the European standard at Zurich in 1898. The Pratt and Whitney Company were also leaders in establishing standards, particularly in screw threads. Professor Weisbach in his 1890 book Mechanics of Engineering that Germany's "Verein deutscher Ingenieure" had attempted to introduce a metric thread system in 1875 but to that time was unsuccessful.

In place of the screw on the earlier machines by 1880 a rack and pinion were installed in the middle of the table. The smaller machines used a conventional pinion while the larger planers used a pinion with V-teeth to gain strength combined with a large circular pitch, while obtaining the smaller machines had smoothness of motion with a small circular pitch and increased numbers of teeth. The gears had cut teeth. The planers were considered wasteful of space, unless parallel with the other machines, and the solution was to make the first transmission from the pulleys to the machine by bevel gears or by a worm gear set.

Borers: At the beginning of the century the engine builders were the major users of machine tools. The works of Maudslay Sons and Field, Boulton and Watt at Soho, and Matthew Murray underwent rapid expansion.

Figure 5-24 Great Western Railway Boring Machine circa 1850

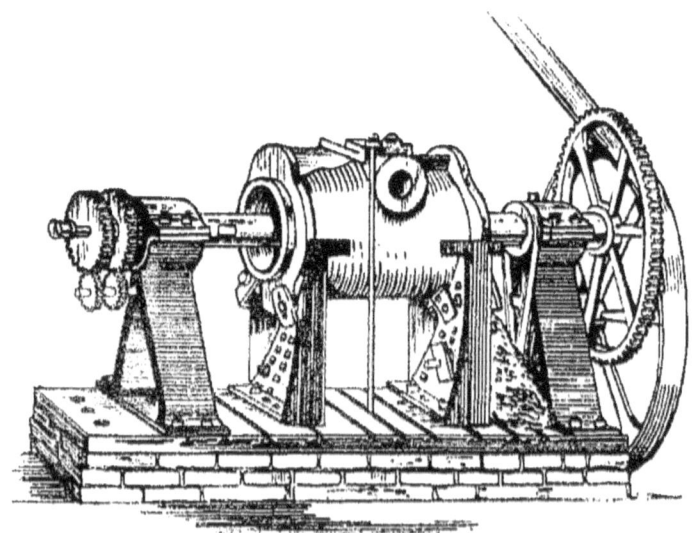

Indicative of the simplicity of most machine tools of the time is the locomotive cylinder borer, installed in the Swindon Works, England, and shown in figure 5-24.

William Murdock, the Soho plant manager, built a massive boring mill based on Wilkinson's machine, but with his own worm and wheel a cylinder was completely machined in 1800 in 27½ days. In 1854 this Soho works built a vertical boring mill that was able to bore the seven foot diameter cylinders for the most powerful engine of the time, the 2,000 hp engine of the "Great Eastern".

Milling: The milling machine was one of the first important U.S. developments in machine tools. The miller replaced single cutting edges with cutters having cutting edges on their circumference. The earliest recorded milling machine is believed to have been installed in Middletown, Connecticut in 1818, and was made by the English gunsmith Robert Johnson. There is also a suggestion that Colonel Simeon North built a milling machine in 1808.

Figure 5-25 Eli Whitney's First Milling Machine

Eli Whitney also made his first milling machine in 1818 as shown in Fig. 5-25, using a worm gear hand control. This machine can be seen in the Sheffield Scientific School, Yale University. The horizontal power driven horizontal table passed at right angles under a rotating cutter. Many believe that the milling machine was more important to the industrial development of the U.S. than the more famous cotton gin.

Between 1819 and 1826 plain and profile milling machines were being made at the Harpers Ferry Arsenal by John Hall. A more advanced milling machine of unknown origin was built two years later followed by a miller in 1835 with improved support for the cutter spindle and vertical adjustment of the headstock made by Gay, Silver and Co.

Milling machines were first built for sale in 1848 by F.W. Howe. Their use was held back by the difficulty of making and sharpening the cutters. Howe was trained in the Gay, Silver plant and joined the Robbins and Lawrence Connecticut factory where in 1847 he built a production milling machine based on designs from the Springfield Armory, the first Universal Milling Machine (Fig. 5-26), and was originally designed to mill the helical grooves in twist drills as well as cutting gear teeth. The cutter was supported and driven between the head and tail centers.

In 1833, David and his son Joseph founded what became Brown and Sharpe, a predominant gear company. They were partnered by Lucien Sharpe in 1850. Joseph would remain the technical leader of the company. J. R. Brown's improvements produced a knee and column universal milling machine with a geared dividing head mounted on a swiveling table.

The No. 12. Brown and Sharpe miller design was known as "Howe's Miller." This universal miller is quite different from the Brown and Sharpe universal miller that followed in 1861. Howe had joined his friend Joseph at Brown and Sharpe to be their first superintendent for the years 1868 to1873. The miller known as Brown's universal miller could replace a number of difficult hand operations. Many of these machines were produced by the company and quickly gained wide acceptance in America. Milling machines would not be seen in Europe until the turn of the century. Elisha Root and Francis Pratt, while building Colt machines at George S. Lincoln Company's Phoenix Ironworks, Hartford, made the machine more compact and robust. The original design was modified with improve rigidity and sold to Enfield in England and further modifications by Pratt and Whitney would make it famously known as" The Lincoln Miller" figure 5-27 which would be sold all over the world.

Figure 5-26 J. R. Brown First Universal Milling Machine 1861

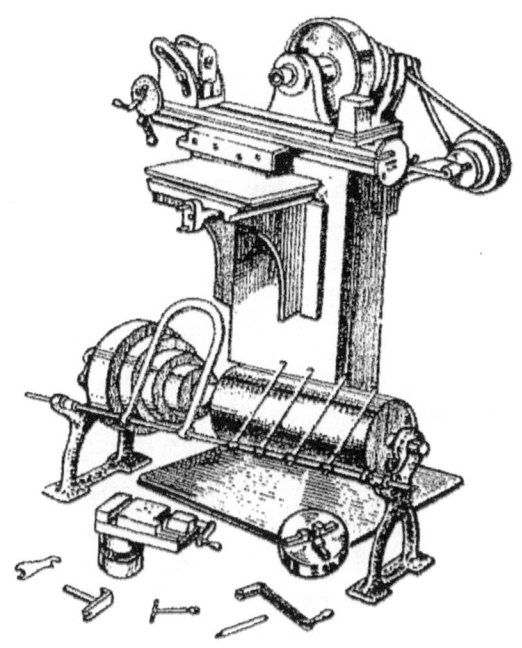

Figure 5-27 Lincoln Miller

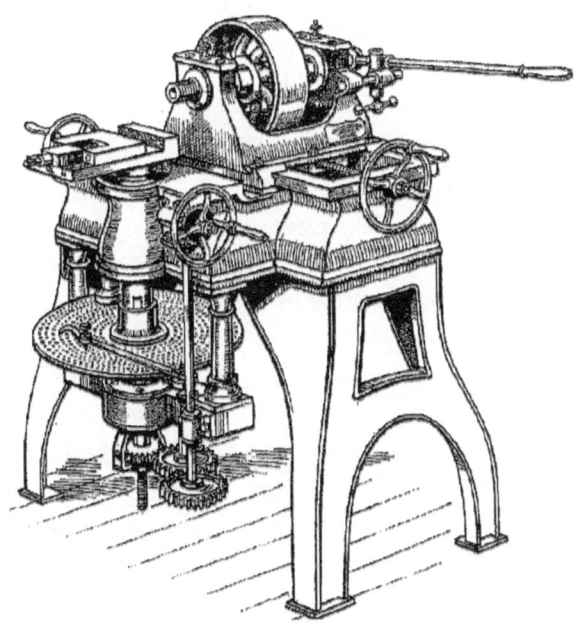

Figure 5-28 Joseph R. Brown's Improved Universal Milling Machine

This milling machine would become the most flexible and widely used machine tool. The feed drive was changed from the chattering rack and pinion to worm and gear and the tailstock was removed. The *"Lincoln Miller"* became popular world- wide and typical orders went to Prussia in 1872 for 72 machines and previously 100 machines to Colt in 1861. At Howe's suggestion the original invention was credited to Joseph Roger Brown in 1861/2 who never claimed to be its inventor. The Lincoln Miller was only suitable for plain milling operations whereas Brown's universal miller (Fig. 5-28) was versatile and needed for short runs and tool rooms. Brown's knowledge of gear cutting persuaded him to include an inclined arbor in the dividing head. In 1876 an over arm was added for support and allowed for the machining of heavier work. The miller was originally for use within the company but when Brown and Sharpe commenced selling tools it was listed in the catalog as the *"Precision Gear Cutter"*.

Joseph Clement, Westmoreland, England manufactured taps and dies using Maudslay's thread standards. Clement introduced the squared shank and is said to be the first to use rotating cutters to cut the flutes (circa 1828). It is of interest because the Cincinnati Milling Machine Company, who would become the largest manufacturer of milling machines in the world, was started by William Holtz in 1880 while making fluted taps in his kitchen on a home-made milling machine.

At the World's Fair in Antwerp in 1893, Darling, Brown and Sharpe exhibited sixteen varieties of milling machines including the original universal machine of 1861. They also exhibited universal grinding machines, gear-cutting machines and screw machines, forty machines in total. In addition they displayed various gear and milling cutters, micrometer calipers and standard gages. Brown and Sharpe also built the first commercial grinding machine in 1864.

Figure 5-29 Joseph Parkinson's 1875 Universal Milling Machine

Shapers: In his plant Maudslay not only built the best lathes of the period, but he was also the best machine tool manufacturer of that time. His major contribution to machine tools was accurate threads, precision and true plane surfaces. A former apprentice to Joseph Bramah, he is credited with establishing thread standards, inventing the metal lathe, a screw cutting lathe and the slide rule. Almost every British machine builder of the Industrial Revolution can trace their work directly or indirectly to Maudslay's plant that also started to build marine engines in 1810. Maudslay was a craftsman with metal and could file a flat surface but not to his exacting standards. He introduced a method whereby the plane surface could be finished with a surface plate, marking compound and hand scraper. A method still taught to European apprentices. The importance of a true flat plane surface cannot be over emphasized.

Joseph Parkinson, Bradford, England built one of the first English universal milling machines in 1875, as shown in figure. 5-29. A former sewing machine salesman, Parkinson exhibited a later model universal milling machine in 1887. This machine was a major improvement over the first machine. He also produced a variety of machine tools and textile machinery.

Sir Marc Isambard Brunel, born near Rouen, France and, an architect in New York City as well as its chief engineer, received a British Government contract in 1800 that required the building of production machinery for the manufacture of naval block-pulleys. Maudslay completed the forty-four machines from Brunel's designs in 1806. He personally simultaneously produced three or four of the surface plates that were initially required using hand tools. Amongst the requirements were a gear driven mortising and a shaping machine. (Figures 5-30 and -31)

Figure 5-30 Maudslay's Production Mortising Machine

Ten blocks were machined at the same time, and indexed by worm gearing. In 1808 130,000 blocks were produced. Ten unskilled men did the work of 110 skilled workers. The figures of metal working machines are from Holtzapffel sketches and show what is similar in design to a modern tool holder for high-speed steel cutters. The machines were built over a six year period. This was the first true example of mass production techniques. The machines were modern in concept, and included a complete range of tools, each performing its part in the planned series of operations.

Figure 5-31 Maudslay's Shaper

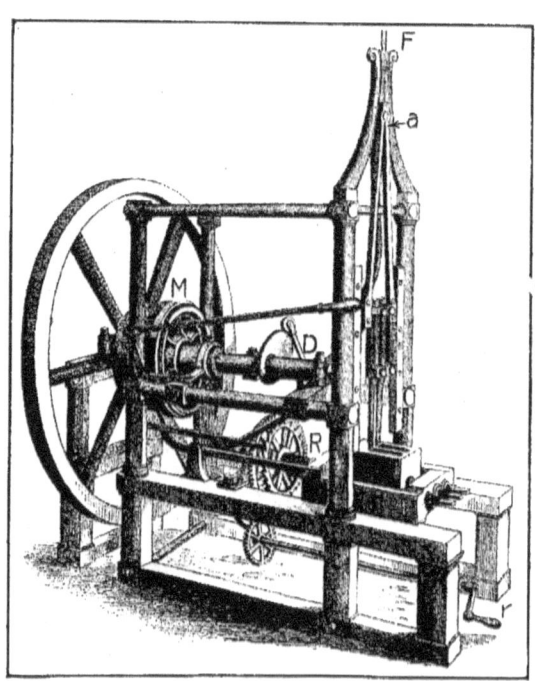

Grinding Wheels: Most grinding tools are in the form of wheels, either grindstones or manufactured abrasive wheels. The early abrasives were of emery, corundum a pure form of emery, artificial corundum, diamond dust, sandstone, and later carborundum. In 1825 corundum was brought to England from India and was later found in N. Carolina,

Georgia and Canada. The corundum was superior to emery but neither could be used in the manner of a natural stone due to their irregular crystals. Until 1860 the granular form corundum was embedded in a softer matrix. Numerous materials were tried as bonding agents with the abrasive with limited success. Solid bond grinding wheels were made in Britain in 1837 were exported to Germany and France.

Vulcanized rubber as a bond for abrasive grinding wheels was patented in Britain by Deplangue in 1857 and manufactured by Warne and Company in London. In 1859 this same kind of grinding wheel was patented in USA by T.J. Mayall. He had two patents #25,747 and 25,841 in '59 and another more advanced patent #125,600 in '72. The wheels were made by New York Belting and Packing Company and sold under the name *Vulcanite*. Their manufacture would continue until 1915 and similar wheels are still used as cut-off wheels.

In 1842, Henry Barclay experimented with vitrified wheels in England. He mixed local clay with emery in moulds and fired the mixture. Barclay was unable to solve the problem of cracking and distortion. Also in England, F. Ransome in 1857, using a lower temperature and a mixture of silicate of potash obtained a more successful soda bond.

In Europe and America in the early part of the century general purpose grinders used either natural sandstones, segments of natural stone; or emery granules embedded in a soft material, wood, soft metal or a leather belt. A high finish could be obtained but without any accuracy. Whitworth's paper read before the British Association for the Advancement of Science in Glasgow in 1840 suggested that hand scraping was superior to grinding as a final finishing operation. He pointed out that the grinding powder was not contollablel. In 1847 Holtzapffel was of a similar opinion stating: *"The entire process of grinding, although apparently good, is so fraught with uncertainty, that accurate mechanicians have long agreed that the less grinding that is employed on rectilinear work the better."*

Nasmyth's disc-type grinding machine built in 1845 revealed the flaws in using a soft metal wheel covered with emery or large sandstone wheels. Previously Bodmer had built a cutter grinder in Manchester, England, using a copper form covered with oil and emery. Nasmyth's machine used a seven foot diameter cast-iron annular wheel with twelve radial compartments. Each compartment was fitted with a fifteen inch stone segment.

The 1880 Census describes the solid emery wheel as an American invention. Small wheels had a velocity of 2,500 fpm and larger wheels as much as 5,000 fpm. Hard rubber or vulcanite was used as the main cementing medium. Some vitrified wheels were run in oil for polishing but the more usual polishing method was to use a leather covered wooden wheel that is then coated in glue sprinkled with emery. The emery wheels were trued and balanced using "black diamond" or "bort". The wheels could be molded and turned to a special profile.

Joshua Rose in his 1890 book "The Complete Practical Machinist" described grinding operations using emery wheels. The natural stone characteristics of Ohio, Nova Scotia, New Castle, Wickersly, Liverpool, Huron, Independence and English were also described, "...to smooth surfaces, to reduce metal to a given thickness, and to sharpen edge tools."

In the U.S. in 1872 Gilbert Hart produced a similar wheel to what Barclay had produced in 1842. Bryan Donkin tested the cutting power and found it to be fifty times better than natural sand-stone. In the following year a potter, Sven Pulson, made an effective clay wheel in the small works of Franklin B. Norton, Worcester, Mass. Barclay's failure was overcome by Pulson on his third attempt. Feldspar was substituted for the slip clay by F.B. Norton and patented in 1877. This was the first grinding wheel that can be considered as truly successful. The grinding wheels started by Norton were made available to the market in 1879, after he had experimented with vitrified emery grinding wheels for four years. Due to ill-health he sold his company in 1885 to Norton Emery Wheel Company, Worcester, Mass., which became the Norton Company. Another American, Charles B. Jacobs was able to make synthetic carborundum in 1897, named alundum. The Norton Company obtained the rights and the new product was the reason for renaming the company. The alundum wheels replaced the natural emery and corundum when it went into production in 1901.

In the U.S. the technical journal "The Manufacturer and Builder" Vol. 7, Issue 9, 1875, reported on a new abrasive manufactured by Van Baerle and Co., Worms, Germany. These stones were made from emery, soluble glass, and petroleum whereas the stones in popular use were made from gum, shellac and emery. When stones of the latter material became heated the gum or shellac became soft and the stone turned greasy. The new stones did not have these problems and could run up to 2,000 rpm and remain cool when polishing steel. The steel would not anneal or have its hardness affected when ground by this new stone which could be supplied in three degrees of fineness, fine, medium, and coarse. The stones were damped by using a petroleum soaked rag.

In 1884 the Detroit Emery Wheel Company carried out a series of tests under the direction of Gilbert Hart. Using emery and corundum grinding wheels the tests proved that corundum was considerably superior to emery. Within the next ten years corundum would replace most emery wheels but because of its scarcity, cost and variable quality a synthetic substitute was sought. A new wheel for cutting arrived in 1888 the "elastic wheel" invented by Henry Richardson, Waltham, Mass.

Not knowing the composition of silicon carbide, by combining the words carbon and corundum, it was given the trade name Carborundum by Edward G. Acheson who commercially developed this new artificial abrasive using an electric furnace in 1891. Excluding diamond it was the hardest abrasive to that time. Silicon carbide had been previously discovered in the laboratory by the French Nobel Prize winner Henri Moissan, developer of the electric furnace. The Carborundum Company was formed in Niagara Falls because they required large amounts of electric power. By 1896 the wheels were commercially available. They were brittle until the density was improved in 1911.

Grinding Machines: All the main types of grinding machine, cylindrical, surface, internal, horizontal, and vertical were in evidence in the late 1850's. However, a visitor to an advanced American machine shop in 1870 would find familiar designs of all modern machine tools except for grinding and broaching machines. Machine grinding started with adaptations of the lathe and would progress into grinders for special duties and applications. The earliest grinders consisted of a lathe with a grinding head and spindle mounted on a slide rest and driven by overhead belts. The hand operated cross slide was used to apply the wheel to the work. Obviously this was a totally unsatisfactory arrangement. When the need for grinding dramatically increased at the end of the century with components required in the millions the lathe method was replaced by the universal grinder. Figure 5-32 illustrates one of these early lathes adapted for grinding.

Figure 5-32 Lathe Grinding Device

J.W. Stone of Washington D.C. was issued a patent for the first surface grinder in 1823. A patent was also issued to Peter Cooper of New York in 1835, which enabled the grinding of flat surfaces. Samuel Darling of Bangor, Maine, patented (No. 9976) in 1853 a surface grinder with present day features. The grinder was in use for many years before and after Darling joined the Brown and Sharpe Company. A patent for an almost identical machine, with the strong possibility it was a copy of Darling's grinder was issued in England to one M. Firth.

In 1833 Jonathan Bridges in the U.S. described the frequently used traverse grinder, traversing the wheel and not the work. James Wheaton of Providence was issued a patent for his Elliptical Spindle Grinder (1834). This patent for a complete cylindrical Universal grinding machine was a major advance in that, by use of a templet form grinding could be achieved. Progress in grinding machines would advance more rapidly in the U.S. than in Germany where Krupp in 1836 built the first all iron cylinder grinder, the largest to date, but they were unable to plane the flat bed.

With a twenty inch grinding wheel the grinder was used for nearly sixty years, but only for simple grinding.

In Scotland in 1838 James Whitelaw built a cylindrical grinder for belt pulleys. The forty-two inch wheel rotated at 180 rpm and the pulley at 130 rpm. In the 1880's Friedrich Fischer in Germany developed high volume manufacture of precision spherical balls at his company Fischer A.G. This was the beginning of the rolling element bearing industry. Until this decade reliable quality bearings were a major problem for gear stability. Frederick W. Lanchester, an English pioneer in automobile design and manufacture, developed important machine tools that included hardened steel roller bearing grinders, accurate to a tolerance of 0.0002 in. The bearings were designed to support his automotive gearing circa 1895.

In 1846 Captain James S. Brown, son of Sylvanus Brown, one of several credited with inventing the slide rest (1791), succeeded Ira Gay in the Pawtucket works that became Pitcher and Brown, specialists in textile machinery.. He became sole owner and enlarged the works in 1847 to four hundred feet in length with over three hundred employees. Captain Brown made many machines of his own design including a bevel gear cutter, boring machine, and grinder. Some of his machines were used for over seventy years.

In 1868 the American J.M. Poole built the first accurate grinder for finishing calendar rolls. Although a modified heavy duty lathe this brilliant machine was reportedly accurate to 0,000025 of an inch using the principle of a pendulum's inertia for the first time in a machine tool. The mechanism acted as a pair of calipers. When the roller is true to size there will be no cut, any inaccuracy would apply pressure to one or both wheels forcing the swing frames out of vertical and grinding until dimensional accuracy is achieved.

A cylindrical grinder for sewing-machine's needles and foot bars was first made as a crude grinding lathe in 1862 by Brown and Sharpe. It was similar to Poole's design and although available for sale, Joseph Brown was well aware of the difficulty in operating the machine. Only one operator, Thomas Goodrum, could operate the grinder and was paid in 1867 the unheard of sum of seven dollars a day. Other than the two owners Goodrum was the only one permitted to wear a silk top hat. To overcome the bottleneck created by this one man operator Joseph designed a "Universal Grinding Machine" in 1868, the prototype would not be built and running until a few days after his passing in July, 1876. A universal grinder was exhibited by Brown and Sharpe at the *Centennial Exposition 1876*. The patent also included a provision for form grinding. Although the writer Joshua Rose described the grinder as a *"Universal Grinding Lathe"* in his *"Modern Machine Shop Practice"* published in 1890 it had little in common with its predecessors. In reverse of past methods the work piece travelled past the wheel.

Henry M. Leland who served his apprenticeship at Brown and Sharpe, before becoming a foreman and stayed there for twenty years, and later was the individual who started the Cadillac and Lincoln Motor Companies, he wrote: *"...I know of none who deserves a higher place, or who has done so much for the modern high standards of American manufacture of interchangeable parts as Joseph Brown."* Leland held several machine tool patents and was in the forefront in developing Detroit's automobile industry. In 1890 he partnered with Charles H. Norton to form the Leland, Faulconer, and Norton Company, which later became the Cadillac Automobile Company, in Detroit to make machine tools, grinders, gear cutters, etc. In 1895 Leland built a grinder for the cycle industry's case-hardened bevel gears, the grinder can be seen in the American Precision Museum, Windsor, Vermont. He was largely responsible for developing the production grinding machines that could take advantage of the new grinding wheels. Leland quickly learned the importance of balancing the wheel and designed a machine to dynamically balance them.

Priot to Norton truing was done infrequently. Norton distinguished himself in the technique of dressing and truing the wheels, insisting that they be trued with a diamond point and not an old piece of wheel as in the past. In 1896 he returned to Brown and Sharpe and developed plunge grinding that made possible form grinding. Meeting severe opposition from the shop superintendent Richmond Viall, Norton returned to the Norton Company which became the Norton Grinding Company with Norton its chief engineer. His first heavy production machine was considerably more robust than any that had gone before, it contained a built-in suds tank with a fifty gpm pump. The machine operator after 1880 could work in tolerances of a thousandth part of an inch, and after 1890 a tenth of one thousandths. In 1869 the regular manufacture of grinding lathes began. Parts of the Putnam lathes (J.W. Putnam, Fitchburg, Mass.), were used to manufacture automatic grinding.

In England, in 1880 Sterne and Co. introduced a gear grinder to rectify cast gears, the first true form grinder for gear teeth. (Fig: 5-33)

Using a template the various teeth irregularities could be corrected. The grinding wheel was trued by hand to the shape of the tooth space. The gear was indexed using its own tooth spacing.

Figure 5-33 Sterne Gear Grinder

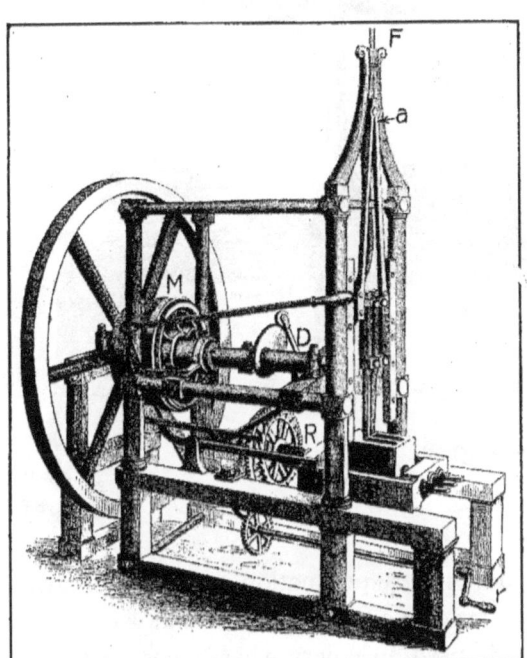

The U.S. Census of 1880 illustrated an early attempt at precision grinding (Fig. 5-34). The modified lathe grinder used an emery-wheel. The belts were driven from a wide pulley on rollers so that the wheel could traverse between the lathe centers. The grinding wheel spindle being driven by a separate counter-shaft. Double bevel gears and a clutch operated the reverse mechanism. The machine was mounted to a wooden floor and ceiling which added to the vibrations. Water was sponged onto the work piece and the guide ways worn by the abrasive mixture. It was next to impossible to prevent chatter.

Figure 5-34 Circa 1860 Grinder Illustration in 1880 Census Report

Faulconer designed a production machine for grinding hardened chainless bicycle bevel gear teeth in 1898 and is credited with being the first to do so. He used an emery wheel whose outer cross section was that of an involute rack tooth. About 1890 Frederick M. Gardner, Beloit, Wisconsin, was mainly responsible for the development of the disc grinder. The coarser grades of emery that had become available also allowed the Gardner Machine Company to build more robust machines. Abrasive ring wheels which were interchangeable with disc wheels brought about wet grinding.

The major importance of the grinder to gear manufacture was the production of accurate hardened steel cutting tools. Milling machines came into wider use and new types of machine tools now entered the market having greater speed and efficiency. The Fellow's gear shaper would not have been possible without precision grinders.

F.R. Hutton's report on the state of the art in machine tools was based on personal visits to the industry centers. All the Eastern states were visited and as far West as Illinois. No southern states were included and the report of power used in iron and steel indicated a fifty percent reduction in the years 1870 -1880 in N. Carolina. This southern state was ranked 23rd in engineering production, S. Carolina was 25th.

Gear Manufacturing: In the first half of the century many of the cast tooth gears were not even approximately the proper shape. Prior to 1830 gears with machine cut teeth would be relatively unknown. Until the 1850's the essential mathematical and geometrical principles of gears were largely misunderstood. By the 1850's production gear making machinery using all the basic methods to produce accurately cut gears became available together with improved tool steels. Those building gear cutting machinery had to take into consideration the type of gear to be cut, the material, the tooling available and if the machine was to cut single or quantity lots. These variables resulted in a wide variety of machines. It was customary, in order to avoid localized heating, to cut one tooth and then move to the opposite side to cut another tooth and so on until all teeth were completed. Prior to the gear generating machine and the modern gear shaper, teeth were cut by machining individual teeth with a form cutter. Each tooth space was a separate machining operation requiring standardized forms to achieve positive uniform motion. Between 1783 AND 1786 Swedish and French physical scientists using the newly developed analytical chemistry realized the various forms of iron were due to the amounts of elemental carbon. The British metallurgists did not take this scientific approach and no technical papers were produced. As the leading machine tool manufacturer their machine tools suffered as a result.

By developing Hooke's (1666) design, James White took out a patent in England for cutting helical gears in 1808. Later in Paris, France, he would build (1824) the first machine to cut helical and spiral bevel gears. It was not a production machine, but could be built to any size. He accurately centered the axes of the dividing plate and gear blank by using bushed bearings. The milling cutter was set at a fixed angle and location, only the blank moved. In his "Mémoire" he stated "je trace et creuse des dents helicon-spirales." He claimed the gears ran perfectly whatever the relationship of the diameters, or angle of the axes. White was a clever engineer but lacked the ability to understand gear tooth interaction.

In 1808 the inventor and engineer Oliver Evans built a machine shop and iron foundry named Mars Works, situated at the corner of Ninth and Vine in Philadelphia. Toothed gears were cast in small diameters of six inch or less, which was most unusual for the time. So called cog-wheels or mortise gear on the other hand would have had inserted teeth of hickory or elm wood. Evans was prepared to supply castings of up to three tons. His personal rules for gear charges were "spurs 5 dolls per foot diameter, bevel wheels 6 dolls per foot. The maker measures to the pitch circle." A notice of 1880 read "...all castings in iron, and wrought iron work for mills and machinery generally, but more particularly iron cog wheels, from the smallest used in machinery for cotton, flax and wool spinning, to the largest used for mills,..."

Figure 5-35 Mortise Gear

By 1821 Richard Roberts had a machine shop in Manchester, England with fourteen machinists operating machine tools to his own design. He developed gages and made extensive use of gears in his designs. His steam powered road vehicle with a differential drive May5th/1821, was the first such application. Holtzappfel credits him with the invention of the slotter and key-seater Roberts has been called the greatest mechanical inventor of the 19th century. As shown in figure 5-36, he re-engineered a clock maker's gear cutter capable of cutting thirty inch gear spur, bevel or worm gears, using a back geared headstock.

Figure 5-36 Richard Roberts Gear Cutting Machine (circa 1831)

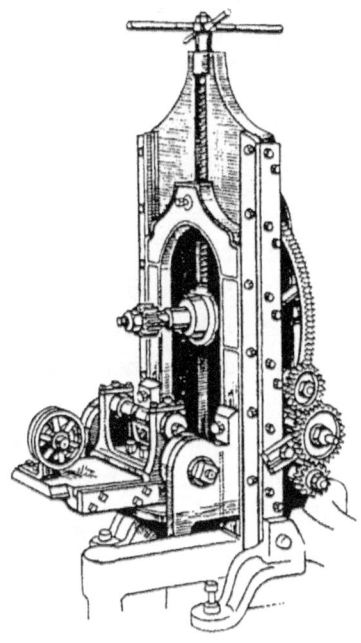

Robert's published in the Manchester Guardian, what is believed to be the first advertisement of its kind: "Respecfully informs Cotton-Spinners, Iron-Founders, Machine-Makers and Mechanics that he has CUTTING-ENGINES at work on his new and improved principle, which are so constructed as to be capable of producing any number of Teeth required: they will cut Bevil, Spur or Worm Gear, of any size and pitch, not exceeding 30 inches diameter, in Wood, Brass, Cast-Iron, Wrought-Iron or Steel, and the Teeth will not require fileing-up."

James Fox was the founder of a well-known firm of machine tool builders in Derby. Fox was one of the outstanding machine tool designers of the time, and a former butler. The transition from crude to precise machine cut gears is said to have started with his machine dated about 1833, which used tooth form cutters with a micrometer screw division of the index cylinder to machine large gear blanks mounted on a vertical spindle. Mounted on a substantial lathe like frame Fox built a production rotary gear cutter. The formed tooth multiple cutters that he used had inserted teeth. The rotary cutter was carried by a slide working in a rigid guide operated by a rack and pinion. Fox also introduced index cutting of gears.

In Waterbury, Connecticut, circa 1830, a new industry had been developed rolling brass into sheets. Chauncey Jerome bought stamped wheels and cut the gear teeth a stack at a time on "perfected gear cutting lathes". Clocks were mass produced and consignments shipped to England as the gears could be relied upon not to warp. "…Three men will take the brass sheet, press out and level under the drop, then cut the teeth and make all the wheels to five thousand clocks in one day. There are from eight to ten of these wheels in every clock…"

In 1833 the Brown and Sharpe Company, Providence, advertised that they will supply "Spur, Spiral and Bevel Geer and screws for worm gear. Dividing plates for all sized Engines graduated in the most perfect manner." In 1855 the company produced their first machine tool specifically designed for cutting gears. Brown's first gear cutter was sold to Providence Machine Tool Company, March 14th 1860. This precision gear cutting machine used a backed off involute cutter invented in 1854 and also used with their Universal Milling Machine.

Whitworth's gear cutters of 1834 were the first machines to use involute formed cutters, geared indexing and cutters driven by a flat belt through a worm gear set. By 1851 the machine had been improved with a self-acting in-feed. The machines had the appearance of a heavy lathe whose headstock carried a vertical spindle to which the involute cutter was fixed. A saddle carried the blank to be cut on a horizontal axis shaft with an indexing wheel.

Figure 5-37 Whitworth's Involute Gear Cutter 1844-51

The Swiss John George Bodmer also patented a gear cutting machine in 1839. This machine was designed to cut internal spur gears, external spur, bevel, worm wheels and racks. However his cutting tools were very difficult to sharpen. His wheel cutting machine could accommodate a fifteen foot gear. He also introduced diametral pitch which would be known for many years as "Manchester Pitch".

In the 1841 third edition of his book "Practical Essays on Millwork" Robert Buchanan illustrated two well engineered gear-cutting machines that were made by F. Lewis, Manchester, England. The machines could cut spur, bevel, skew-bevel, and worm wheels up to five feet diameter in iron, and ten feet diameter in wood, widths to fourteen inches. Any pitch or number of teeth were obtained by changing the number of teeth in the worm index The gear cutting machines bore a similarity to milling machines, as can be seen in figure 5-38, a "Millwork" illustration.

Figure 5-38 Lewis's Gear Tooth Cutting Machine

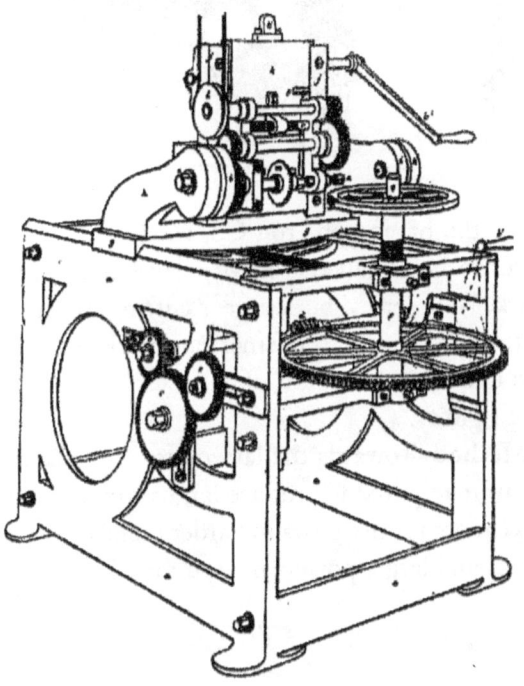

Figure 5-39 Gould Gear – Cutting Machine 1855

Bement, for the St. Joseph Iron Co. Mishawaka, Indiana, built the first gear cutter to be installed west of Cleveland, Ohio in 1846. Ezra Gould, Newark, New Jersey, built a gear cutting machine in 1855. This machine is in the possession of the American Precision Museum, Windsor, Vermont. The four legs are typical of these early machines built with an obvious lack of rigidity.

Milling Method: Towards the latter part of the century all types of machine tools were required to produce large production runs. A pressing need to produce more precise gears in harder metals arose from the higher rotational speeds and larger powers now coming into use.

A Gear Chronology

Figure 5-40 Standard 1880 Bevel Gear Machine

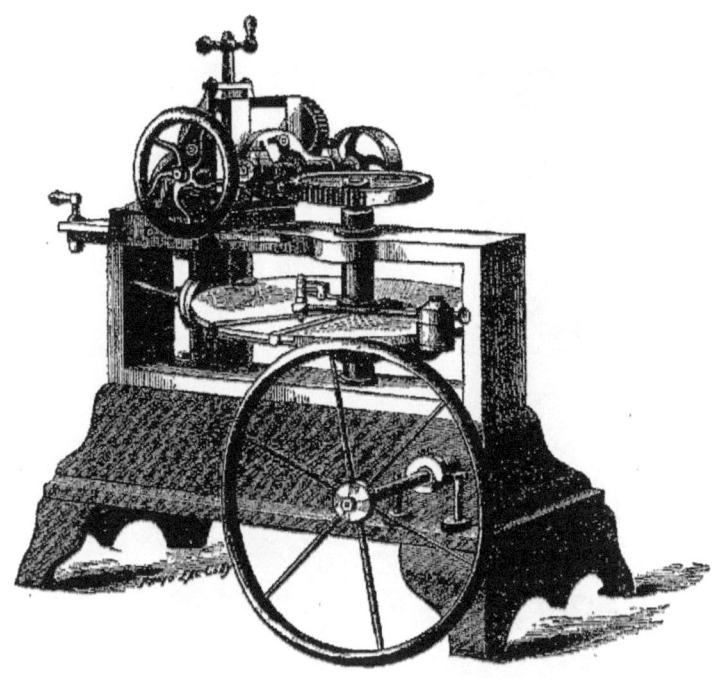

Fifteen individual gear cutting machines were illustrated in the industrial census of 1880. Figure 5-40 illustrates what was termed a standard machine. The cutter carriage could be swung and set to cut bevels of any angle. The cutter was automatically fed across the face. After 1850 other large gear cutting machines were developed for engines and power transmission.

Figure 5-41 Milling Machines Adapted for Gear Cutting

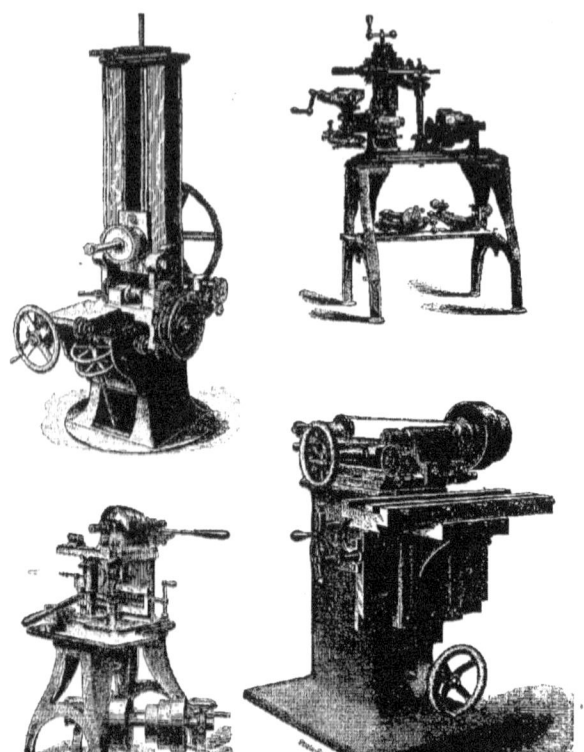

Machine tools were regularly adapted to produce these higher quality gears as illustrated in figure 5-41 shown in the 1880 Census Report on Machinery. The first figure in the illustrations was the basic gear cutting machine, the figure top right and bottom left were adaptations to cut bevel gears. The machine on the lower left was designed to cut racks. Special millers were also being designed to cut threads.

Following Brown and Sharpe's introduction of their formed cutter the principal method of manufacturing cut tooth gears, until the latter end of the 19th century, was with the milling machine. Universal millers could be used to cut gear teeth with a universal head that had a worm wheel and index. To increase the adaptability of the milling machine by 1880 several modifications had been produced, especially for cutting bevel gears. The miller shown in figure 5-42 was patented in 1897 and was awarded the gold medal at the 1900 Paris International Exhibition. It enabled more accurate worm gears to be made. After twenty-five years service the Holroyd thread miller was put on display at the South Kensington Museum.

Figure 5-42 Holroyd Thread Miller

Odontograph Method: Known as the third Translator, Professor Robert Willis (1800-1875), a Cambridge Professor of Natural Philosophy, provided numerical tolerances and invented the first odontograph (circa 1840).

Figure 5-43 Professor Willis's and Dr. Robinson's Odontographs

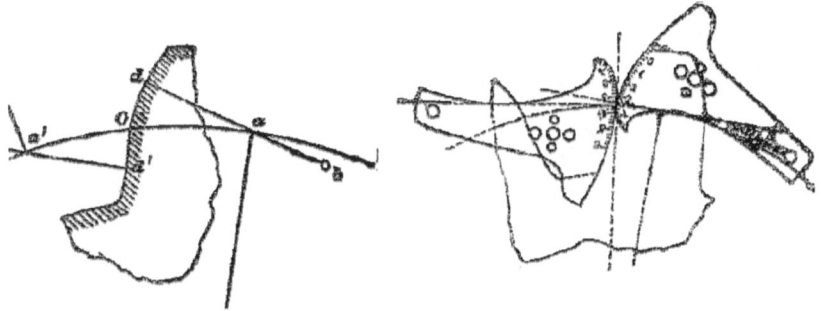

A three point odontograph was a simple method of recording the positions of the centers of the circles which were the closest approximation to the desired tooth curve. Willis's odontograph was widely used for drawing tooth profiles and calculating the distances from the centers

of curvature for tooth faces and flanks for various pitches and different numbers of teeth.

Holtzapffel and Company advertised the odontograph in 1843 as follows: "THE ODONTOGRAPH, Invented by the Rev. R. Willis, A.M., F.R.S., Jacksonian Professor, Cambridge. This is an instrument of easy application, used for describing the teeth by circular arcs, so that any two wheels of a set may work together. Price of the Odontograph on card and varnished, 5s. The theoretical explanation of this system of teeth, which has been extensively adopted by practical use will be found in the Trans. Inst. Civil Engineers, vol. 11., and in Willis's Principles of Mechanism, 1842."

By 1850 for large or single gears the preferred method was the odontographic or templet method using a shaping or single point tool. The method required the tool, as it was fed into the blank, to follow a master templet with the correct curvature. The alternative method available was using a fixture on the tool head that applied a radial motion to the tool. Single point cutters and rotary cutters with the required tooth profile could be used in the same way. Both methods could be applied to the shaping machine using a single point tool. The alternative was using a tool whose cutting faces were in the shape of the tooth profile and by 1850 all these methods would be in use.

Mac Cord also built an advanced Odontoscope as shown in figure 5-44. The wide acceptance of the method is illustrated by its inclusion in text books, as recently as 1927. Involute and cycloidal odontograph tables were taken from Grant's "Treatise on Gear Wheels".

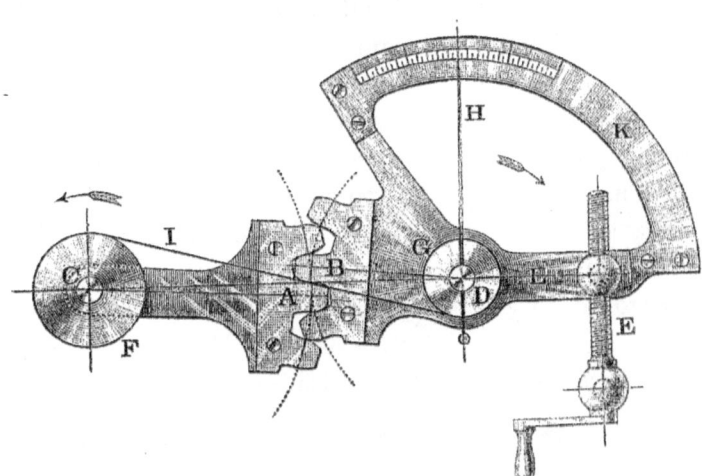

Figure 5-44 Mac Cord's Odontoscope

Professor J.F. Klein in his 1889 book "Elements of Machine Design" wrote that the general practice to determine tooth profiles was by using templets to make a wooden pattern or circular cutter. This method however is inaccurate for bevels gear as the spherical cycloid or spherical involute are non-developable.

Planing-Generating: The significant difference between the formed tooth cutting of individual teeth and generating is that by in the latter the teeth are conjugate related. The motion between the gear blank and the tool is positive and uniform during the generation. When teeth are cut individually the teeth require proportioning for uniform motion. It has been said that in the new methods development is towards the objective and in the other is backwards.

In France (1829), Glavet was the first to patent a templet gear planer that was capable of cutting heavy mill gearing. The machine, with a single point tool, had a reciproating shaper like action. The gear planer may be even older.

George H. Corliss, Providence, R.I. engineer and inventor first patented his templet planer for bevel and larger mill gears in 1849, and was the first in the U.S. to use this type of gear planer design. Considered to be the most important trade show ever held the 1876 Centennial Exhibition exhibited 8,000 operating machines in the Machinery Hall. Thousands of state of the art machine tools were sold. The exhibits were powered through five miles of line shafts driven by a 1,400 horsepower Corliss steam engine. The engine is said to have worked quietly and with less vibration than a watch. Later sold to the Pullman Company, and moved to the South Side of Chicago, the engine supplied power to their plant for the next thirty years. The 70-foot tall steam engine was the largest ever built to that time. The two walking beams were connected to a bevel gear between the cylinders. This gear was the heaviest cut gear to date. The gear weighed 56 tons and had a 24-inch face, and exceeded six feet in diameter. The flywheel gear was thirty feet in diameter. Also on display at the Exhibition was the planer illustrated in figure 5-45 that cut the engine's huge bevel gears. Prior to this they were inaccurately produced on milling machines.

Figure 5-45 Corliss Bevel-Gear –Cutting-Machine

The gear "planed out, element by element, with absolute theoretical precision." The cutting point of the tool is made to always travel in a right line passing through the vortex of the pitch cone. A larger scale template reduced unavoidable errors and an outline accurately conformed to that of the tooth's transverse section. Bevels and spur gears up to one hundred inches in diameter could be cut. The index wheel was built fifteen feet in diameter with the intention of reducing errors. The former guided the tool so the cutting point to each element of every tooth will pass through the apex and be tangent to an exact profile at the base of the cone. A smaller machine was also built by the same company.

Figure 5-46 Layout Corliss Machine

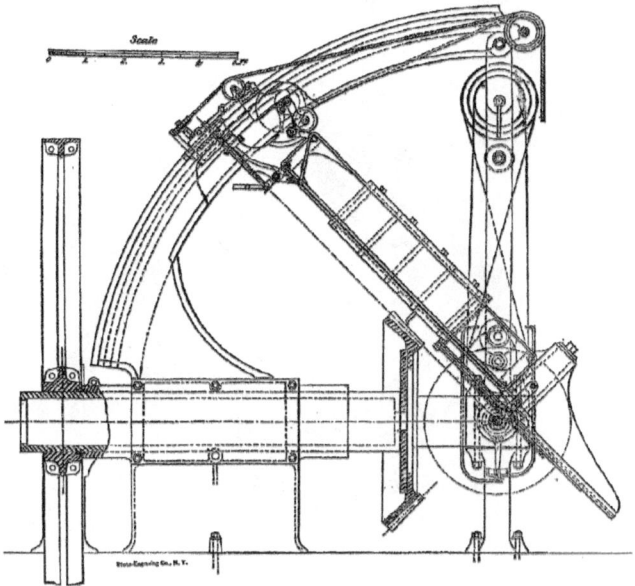

An edition of "The Engineer" the predecessor of "Power Engineering" was issued on June16th, 1910. The cover celebrated the centenary of the steam engine with an article on the Corliss engine and a picture of R.M. Corliss. The cover was reproduced to celebrate the journal's second century in their June, 2010 issue.

In 1865 the Irish born William Gleason founded the Gleason Works in the U.S. He built a template machine to make spur gears up to twenty foot diameter. Gleason had learned his engineering skills at Colt Armory and Pratt and Whitney, as did many other successful tool builders, such as Bullard, Lapointe, Gardner, Bardons, and Foote.

William Gleason had invented the first commercially practical method for machining straight bevel gears in 1874, based on the planing principles shown in figure 5-47.

Figure 5-47 Principle Gleason Planer

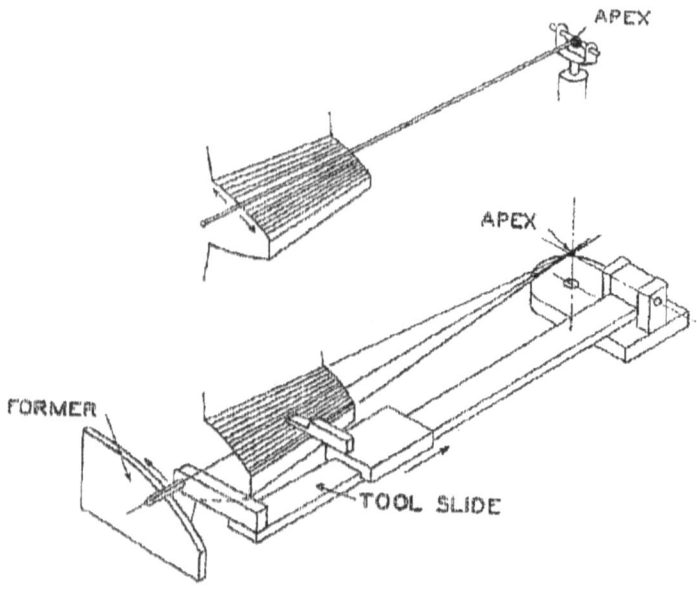

Figure 5-48 Gleason's 1874 Planer Based on Original Patent

The machine illustrated in figure 5-48 was patented in 1876 and used for forty years before being donated to the Ford museum, Dearborn,

Michigan. It was the first commercially practical machine. The teeth were straight and directed towards the apex. The shape of the teeth were copied from a template that was an enlarged version of the teeth to be cut.

Figure 5-49 Gleason Bevel/Spur Tooth Cutter circa 1875

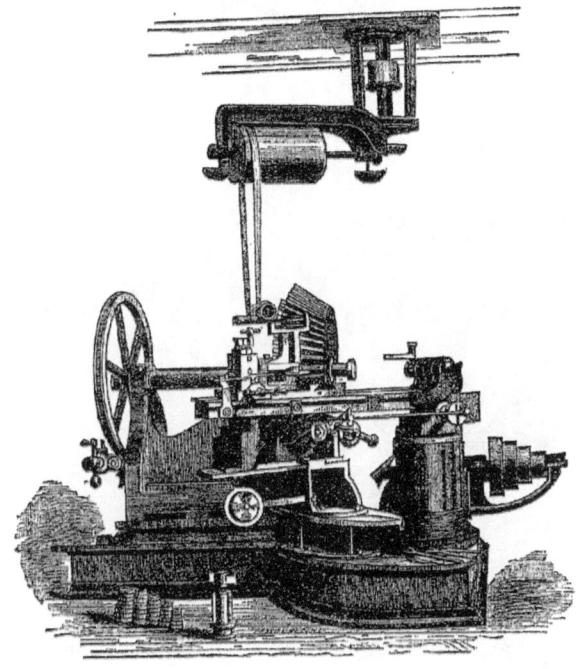

A similar machine to the 1875 Gleason shown in figure 5-49 was built in the U.S. by the Holmes Company and is illustrated in figure 5-50.

Figure 5-50 Holmes Bevel Tooth Machine

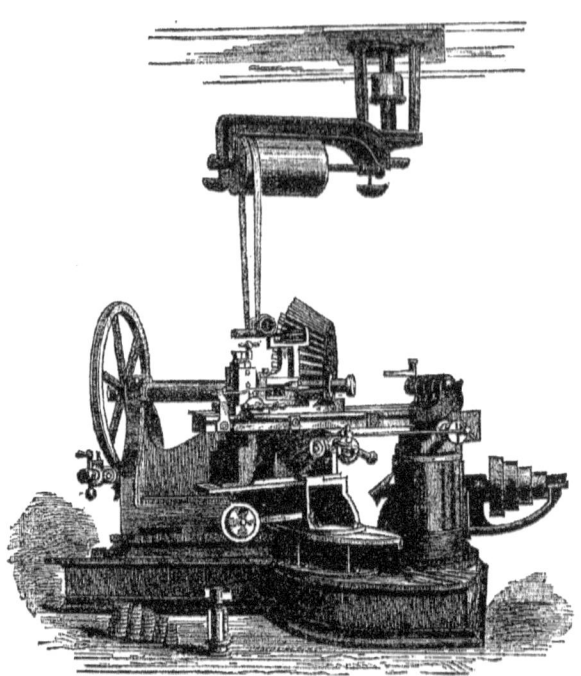

In the 1870's the Gleason Works were building gear cutting machines with increased capacity as shown in figure 5-49. The new machines could cut spur or bevel gears up to one hundred inches in diameter. The tooth former is placed under the tool holder which is fed above it. The radial box will swing to any angle with the mandrel between 6° for spur gears and 90° for crown wheels. It is also hinged to allow bevel gears to have vertical movement.

In Germany in 1877 Professor Julius Herrmann published "Verhandlungen des Vereins zur Beförderung des Gewerbfleisses" in which he provided machine tool modifications for the planing of bevel teeth. The general planing method had previously appeared in Weisbach-Herrmann's "Machinery of Transmission". Dr. Weisbach had written a full series of text books that were published in 1890, "Mechanics of Engineering and of Machinery", the 2nd edition was revised by Gustav Herrman and translated by Professor J.F. Klein at Lehigh University. Volume 111, "Mechanics of the Machinery or Transmission" contained numerous formulae.

Hugo Bilgram was to develop a modification of Herrmann's planing method in 1883. The new and exact method by Bilgram was described in the Journal of the Franklin Institute August, 1886. Gear cutting machines are said to either mill or plane the teeth. In milling the cutter is rotating and in planing the cutter has a straight line reciprocal motion. When the blank is moved as in mating with another gear's tooth the teeth are generated. The planer cuts the groove in accordance with its outline; the disc milling cutter and formed end-mill interfere with the sides as the teeth are entered. The development was driven by the chainless bicycle.

Figure 5-51 Sunderland Gear Planer

The 1895 gear planer shown in figure 5-51used rack type cutters and was developed by Sam Sunderland, Keighley, England. This reciprocating method (termed molding-generating or plano-generating) was so successful that the planer was still being manufactured in 1983. The spur gear generator uses a rack with cutting edges that is reciprocated parallel to the blank axis. The cutter box moves to and fro across the whold face of the blank, and the slide which guides it is given a slow vertical movement in time with the blank's rotation. The cutter has a downward movement through a distance equal to the pitch of the teeth the blank turns through an angle equal to one revolution divided by the number of teeth.

John Comley in the U.S. attempted to roll gear teeth on heated blanks by a process similar to knurling. There were similar attempts with rolling gears elsewhere but all would be unsuccessful until the achievement of H.N. Anderson in 1911.

Describing Generating: From the mid 1800's until 1884 all production machines worked on the describing-generating principle. The method varied from molding-generating only in the form of tool used. A theoretical point was used for describing-generating and a portion of mating rack with a cutting edge formed into the side of a tooth for molding-generating.

Joseph Saxton of Philadelphia and later in London invented a new type of generating machine to cut watch wheel teeth to the true *epicycloidal* form in 1842. Although they were clock gears, it is a first in gears cut on the describing--generating principle and designed by an American. This was the most amazing advance in gear cutting. Hawkins described Saxton."... *who is justly celebrated for his excessively acute feeling of the nature and value of accuracy in mechanisms; and is reputed not to be excelled by any man in Europe or America for exquisite nicety of workmanship."* This method of correctly generating curved teeth and truing them after they had been roughly formed can be seen in Fig. 5-52. The method ensures that the faces of the teeth are milled true. The cutter lies in a plane through the axis of a describing circle, which rolls around a pitch circle clamped to the blank. The serrations in the two wheels that connected the two shafts were milled with very fine teeth so that they rolled together in the same manner that as the finished gears This was considered to be the most accurate method pre 1850.

Cycloidal tooth forms in general could not be produced with a formed cutter. Saxton produced a straight-sided milling cutter to generate an involute tooth, establishing the generating principle, and the first generating machine of which we are aware. The faces of the teeth were milled true, the side of which lay in a plane through the axis of a describing circle which was rolled around a pitch circle clamped to the side of the gear being cut. These circles were gear wheels milled with fine pitch teeth. The teeth are generated by rolling the cutter around the blank at the correct radius. Gear cutting machines with indexing arbors had been designed in England and used or misused to make any operator's idea of form or pitch. Saxton's machine was the first that worked on the describing generating principle, and required an engineer who understood gear theory.

A Gear Chronology

Figure 5-52 Saxton's Gear Generator

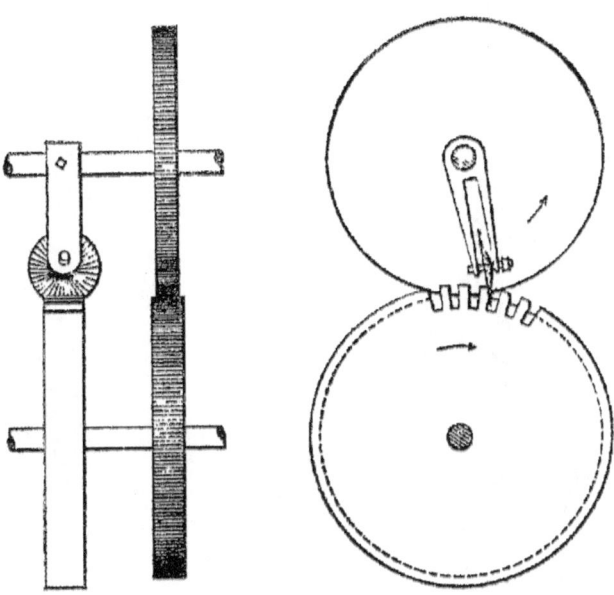

Instead of an indexed arbor on the work table in generating the gear blank rotates continuously synchronized with the cutter. In practice the cutter operates as a master gear reproducing itself in the blank. Because of its complexity the method was rarely used prior to 1850.

> A typical early describing generating tool was described in the British patent No. 2714 that was issued to Lawson and Cotton in 1843. While ingenious it could not produce the hypocycloidal flanks needed for interchangeable epicyclic gears. This method was first practically applied by German engineers E. Hagen-Thorn in 1872 and Gustav Herrmann in 1877. Hagen-Thorn did not build a machine but illustrated how a competent engineer could adapt a shaping machine to cut involute gears using a double-pointed tool for cutting each flank separately. Hagen-Thorn wrote in 1872 on how to generate involute teeth in both spur and bevel gears. The method was further improved after careful Figure 5-45 Corliss Bevel-Gear –Cutting-Machine.

A theoretical analysis by Herrmann in 1877. To quote Grant's comments on the gear literature of his day: "...Professor Herrmann's section of Professor Weisbach's book "Mechanics of Engineering and Machinery" is the most important work that can be named..."

Figure 5-53 Form Cut Accuracies.

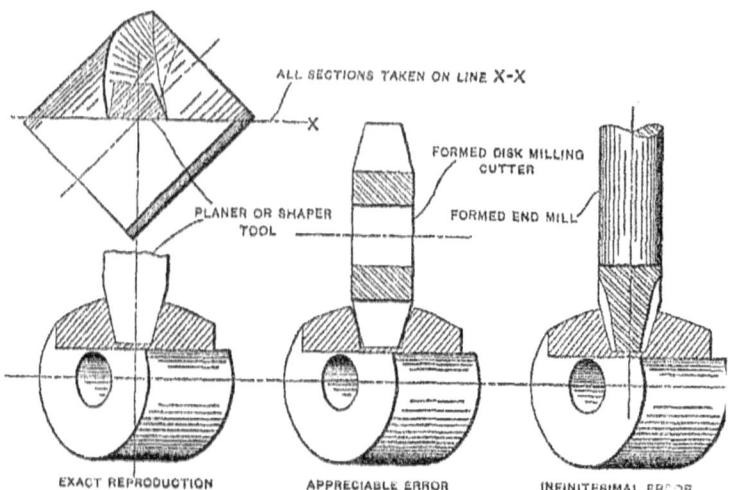

Herrmann's method was an even easier method of adapting a shaper by use of a single point reciprocating tool similar to the design of future shapers. The blank motions were controlled by various attachments so spur or bevel gears with either involute or epicycloidal teeth could be produced. A similar machine for cutting epicycloidal teeth in bevel gears was patented in Vienna by C. Dengg in 1879. Pratt and Whitney produced an epicycloidal machine in 1880 used a circular cutter that was controlled by a disc that rolled on the outside of a ring that represented the pitch circle of the gear being cut.

Mr. Ambrose Swasey, Cleveland, Ohio, invented the Swasey Method in 1889. At the time this was considered to be the only practical application of the describing-generating process. He used a small cylindrical cutter in a machine similar to a lathe. It was in use for the next twenty years at the Pratt and Whitney Company for making gear tooth cutters, before the day of the formed cutter. Pratt and Whitney also made the machine available for sale. Grant also designed machines of this type. The method was more popular in Britain than in the U.S. because though accurate it was slow and more suited to short run production.

A Robey-Smith machine designed by J. Buck in 1895 used two cutters working on opposite sides of the tooth. The machine is on exhibit at the London Science Museum. The machine was exhibited by Smith and Coventry with their own describing generating machine at the 1900 Paris Exhibition.

Molding-Generating: The application of a worm sectioned to provide cutting edges was first used by Ramsden in 1768. The theoretical basis for all gear generating machines was provided by the Scotsman, Professor Edward Sang in 1852. Prior to Sang's work gear generators were being built on the wrong principles. Molding-generating either a straight toothed rack or a gear shaped cutter is used. Sang was the first to make the transition from theory to practice, reversing the usual processing order, and fully developing gear theory. Sang realized the importance of the path of contact, and stated that the form of that path was dependent on the tooth contour. The term odontoid was used for the optimum tooth curve. His elaborate treatise "New General Theory of the Teeth of Wheels" made an outstanding contribution in advancing the gear art. The analysis indicated a rack tooth shaped cutter could be used in an intermeshing method to make interchangeable gears. It was Sang's Theory that demonstrated that a screw could be employed to cut its own wheel. In cutting a guide template the screw automatically finds the form of a tooth conjugate to that of a given rack. The screw thread may be of any desired form. Pratt and Whitney's pantagraphic gear cutter was based on this principle. Mac Cord in his 1883 book critiqued Sang and wrote: "His treatment of the subject is analytical and abstruse, the style obscure and pedantic to the last degree."

Hobbing can be considered as a development of the milling process. A gear-shaper uses a cutter whose teeth are the same size and shape as the teeth of the mating gear. The angular relationship and alignment of the axes and the centers of the cutters and blanks are critical to its success, and at the same time difficult to determine. The hob also requires precision grinding in order to maintain its accuracy.

The molding-generating, or hobbing, as we know it today, which used a worm like hob, progressively cutting the tooth space as the tool and blank rotate in sync. The hobbing process is carried out on a machine tool we know as a hobber. An action is generated between the hob and the work on a continuous basis, similar to a rack and gear in mesh, the rack having a forward and backward motion.

Apart from the Fellow's method this created a new problem as all other machines using a single point tool or rotary cutter to cut one tooth at a time. This method created localized heating and deformation of the finished gear. The problem was partially ameliorated by not cutting the teeth consecutively. Bodmer to our knowledge was the first to use the molding-generating principle. One of Bodmer's machines, some were still in use in 1904, had attachments for making internal gears, worm wheels and racks. His worm wheel cutter, was identical with the worm with which the worm wheel to be cut is to mesh, and made with cutting teeth. Bodmer was the first engineer to use a true form milling cutter to produce metal gears. His cutters were made on a gear cutting machine fitted with a milling cutter and a templet. Wooden patterns for gears were fly-cut. The successful application of hobbing was complex and despite the efforts of Bodmer, Whitworth and Schiele in 1856 and others issuing numerous patents it was not until 1887 that a fully geared machine was produced, the work of Grant.

Bodmer patented many machines and had them drawn to detail. He also equipped a machine-tool plant so that the tools and layout would economize on labor, movement, and energy. This Manchester, England, machine shop was a model of how such shops should be arranged. Other than his gear cutting machines his machine tools for the most part were unremarkable. His patent in 1839 for a vertical facing lathe was not accepted in Britain but was later developed by E. P. Bullard and Conrad Conradson in the U.S. in a multi-spindle fully automatic form. If the hobber was rigid, full advantage could be taken of the high-speed steel cutters to produce precision gears at a previously unheard of rate without heat distortion.

Hugo Bilgram, in Philadelphia, Pa., invented the first successful hobber that was patented in the U.S. (# 294,844) in 1884. Several hobbing patents were issued in the U.S. in the latter half of the century. In 1885 an original hobbing-machine patent was applied for in the U.S. Swasey built an advanced machine in 1885 for cutting spur gears on the molding-generating principle. The rotating milling cutter passed across the blank's circumference, their axes being at right angles to one another. The synchronous motion produced the effect of a straight rack and gear in mesh. A U.S. patent 405,080 was issued June 11th 1889, stating all gears will be interchangeable if cut with the same cutter. U. S. Patent No. 20,023 was issued to Henry Belfield for a lathe hobbing attachment in 1871. The required accuracy of the hob was not possible without the advances made in precision grinding.

The first fully geared hobber was patented by George B. Grant in 1887. Jüngst J. E. Reinecker in Germany, Frederick Lanchester in England manufactured gear hobbers in 1893, 1894 and 1895 respectively. Lanchester's hobber can be seen in the Birmingham Museum and Art Gallery, England. He also built the first British hobbing machines designed for worm gear production. His automobiles also contained epicyclic gears, and were the first to have a differential in the rear axle. Hermann Pfauter Germany produced a specialized hobber in 1897, the first universal hobber. The hobber could produce helical and spur gears with a hob that could work at any angle. Gear hobbers would be slow to develop until the machines of Jüngst and Heinecker introduced their horizontal spur gear hobbing machines in 1893. They also supplied the hobs. This slow development was probably due to design inadequacies.

J. E. Reinecker, Chemnitz-Gablenz, Germany, built a machine for cutting gears by the formed milling process. The cutter was driven by a large diameter internal gear mounted to a swivel table. Unlike the usual milling machine the work spindle was at the top of the column and the cutter spindle on the knee. The differential was used to effect the combination of the helical and indexing movements. Helical gears for parallel or crossed axes and worms can be hobbed by either differential or non-differential methods. When using the differential method to cut parallel shaft helicals, the feed is disconnected, and the lead relationship of the hob and blank is unaffected. This is an advantage when two or more cuts on the same machine were required.

Figure 5-54 Section Showing Differential Reinecker Universal Gear Machine

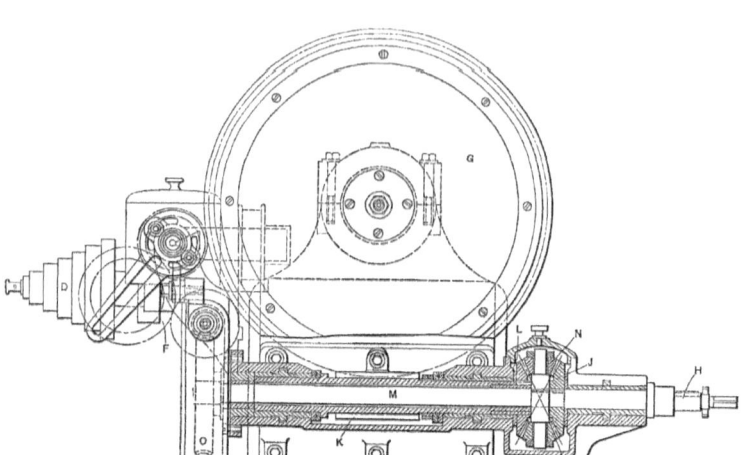

As shown in figure 5-54 the indexing was by the worm wheel G which with gears J, L and N forming a differential mechanism. These early machines had learned that it was important to place the differential in advance of the change gear train. There were other advantages but for parallel helical gears the disadvantage lay with the limited number of helix angles, pitches and number of teeth.

The first use as a production tool was by Joseph Whitworth in 1835 when as a "hob" to cut spiral gears. Whitworth also patented a worm gear hobber (British Patent No. 6850) in 1835. The gear blank was geared to the hob through a gear train, generating for the first time involute gears commercially. His machines were rigid and ahead of their time. By 1851 this basic machine had been improved with a self-acting in-feed. The involute cutter was mounted to the headstock's vertical spindle and had the appearance of a heavy rigid lathe. The gear to be cut was mounted on a horizontal axis shaft that was indexed. Whitworth's hobbers remained in worldwide use without major modifications for sixty years. These were the first machines with involute cutters, geared indexing, and power-driven through a flat belt and worm gear drive. Grant in his 1899 treatise commenting on Whitworth's hobber "...invented in1835, and has not materially improved since then." In fact, by 1844 Whitworth had perfected a huge machine to cut involute gears using a formed milling cutter. The machine was fitted with geared indexing for the blank arbor and a cutter power driven through a flat belt, worm and wheel. Further improvements in 1851 included a power feed to the cutter head.

The modern gear hobber was formulated by Whitworth and Christian Schiele. The

British patent No. 2896 in 1856 was for Schiele's machine designed to cut spur, helical and worm gears using a hob at 90°. The machine used gears to rotate the blank and hob at relative speeds. This was the first hobber to use this principle. It had been recognized since Ramsden in 1768 that the worm is a form of continuous rack that can cut gear teeth if it is provided with a cutting edge.

Figure 5-55 Grant's Conjugator

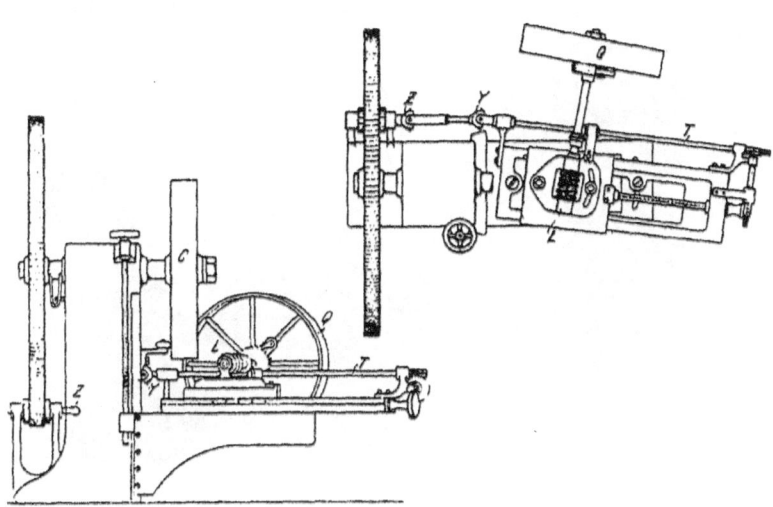

Grant patented a hobber (#407,437), and previously the *Conjugator* (patent #405,080), an extension of the Whitworth hobber of 1835, that could hob spur or helical gears. (Fig:5-55) He used a describing method to generate tooth profiles. A single point tool reciprocated along a guide that emulated a cone which solved the problem of coordination between the revolving work-piece, the feed, and tool rotation. He described the hobbing process for a worm gear: *"If this cutting spiral gear* (hob) *is mounted in connection with an uncut blank so that both are rotated with the proper speeds, and the shafts of the two gears are gradually brought together while they are revolving, the edge of the blank will be formed with concave teeth that curve upwards about the sides of the cutting gear."*

Molding: The first true gear-molding machine was patented in Prussia by J.G. Hoffman. This method replaced the old method of molding gears from patterns. Machine molding provided teeth as accurate as machine cut teeth if the uneven shrinking of the casting was avoided. All types of gear can be molded by this process. A "cheek" is used to ram up a gear segment and then the segment is moved around and another set rammed up.

Figure 5-56 Mesta Gear-Molding Machine

George Mesta founder of Mesta Machine, Pittsburgh, developed and patented three sizes of molding machines the largest of which had a maximum diameter of 156 inches. All the teeth were molded from the same pattern or tooth block, as shown in figure 5-56. The sand was packed into the tooth block and supported by pieces of wire. After being swabbed with plumbago in solution with syrup water and drying, the mold is closed and then baked in large natural gas furnaces.

Gear Shapers: In 1884, Bilgram built a gear shaper applying the molding-generating principle for machining the small bevel gears that were a necessary part of the current bicycle designs. The shaper also provided a new tooth form, the *Octoid*. When he was twenty-two E.R. Fellows, a window dresser with natural engineering ability, was brought by James Hartness into the Jones and Lamson Company in 1889. Such was his success he founded the Fellows Gear Shaper Company, Springfield, Vermont, introducing their gear shaping method in 1895. This gear shaper shown in figure 5-57 used the molding-generating principle. The shaper greatly facilitated the making of internal gears. This method generated gears with a fully formed pinion type cutter that had a backed off cutting edge and a hardened and ground involute profile. This was a vastly different tool to that of the straight rack principle used by Swasey. The machine was similar to a vertical slotter, with the cutting tool and gear blank rotating in mesh. The Fellow's method produced accurately generated involute teeth. The design was a radical departure from the use of rotary cutters. The cutter and blank were held on parallel shafts. One cutter of any pitch can cut any number of teeth of the same pitch. The original gear shaper cutter was the first accurately generated gear cutting tool ground all over after hardening. The cutter action is shown in figure 5-58. Within two years Fellow's invented a gear shaper, gear-cutter grinding machine, a Figure 5-57 Fellows 1895 Gear Shaper bevel gear generator as shown in figure 5-59, and the vertical gear shaper that was to make his company famous.

Figure 5-58 Fellows Cutter Action and 1897 Cutter-Grinder

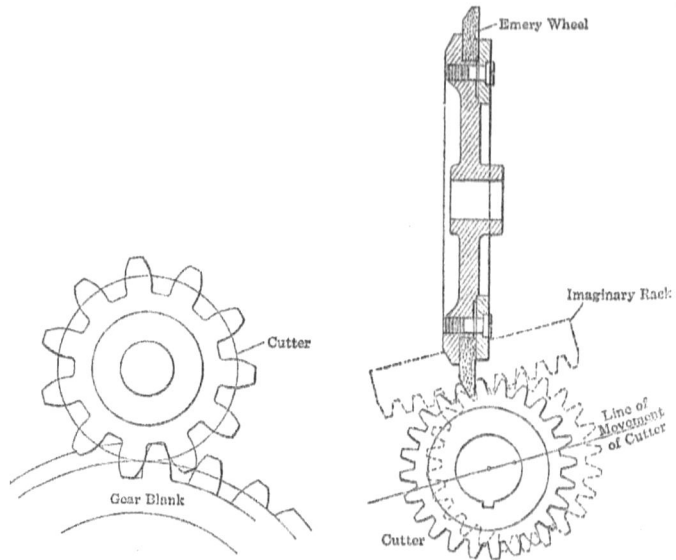

Cutter Action Generating cutter teeth

Figure 5-59 Fellows 1897 Gear-Cutter Grinder and Bevel-Gear Generator.

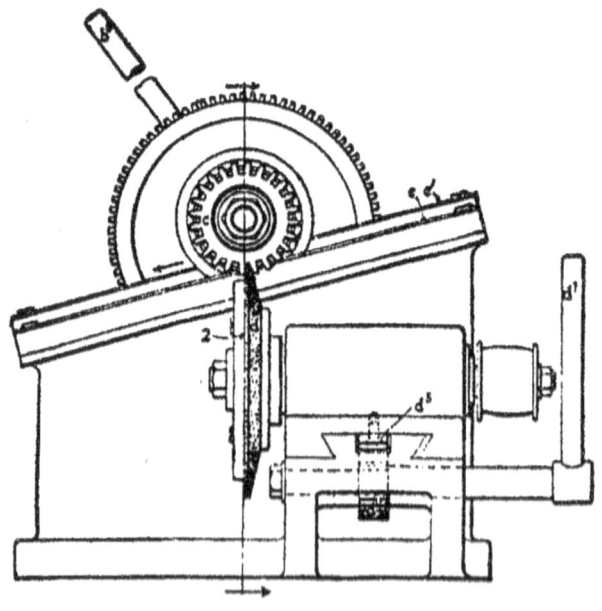

Bevel Gear Machines: Bilgram's bevel gear generator was originally built to specifically produce bevel gears for the then current chainless bicycle demand. Unlike the machines that followed, it used a single point cutter. He produced a graphical method for 10th 1883. The Octoid bevel tooth form and an ingenious machine for planing the form are also Bilgtam inventions. Both, the octoid tooth form and machine, were described in the American Machinist May 9th 1885, and the Journal of the Franklin Institute, August 1886. The hand operated machine was in popular use for many years before Bilgram developed an automatic version on the same principle as the generator shown in figure 5-60.

He produced a graphical method for determining bevel gear dimensions, published in the American Machinist on November 10th 1883. The Octoid bevel tooth form and an ingenious machine for planing the form are also Bilgtam inventions. Both, the octoid tooth form and machine, were described in the American Machinist May 9th 1885, and the

The odontoid, or pure tooth curve, was defined for the contact of the pitch lines at the pitch point in a continuous and progressive manner. In 1900 Bilgram made the bevel gear machine automatic and was awarded the Franklin Institute gold medal. The same planer would be used to cut a later invention, the planoid bevel tooth. He also modified the machine to cut spur and helical gears. Some of his machines were built under license by

5-60 Bilgram's Automatic Bevel Machine.

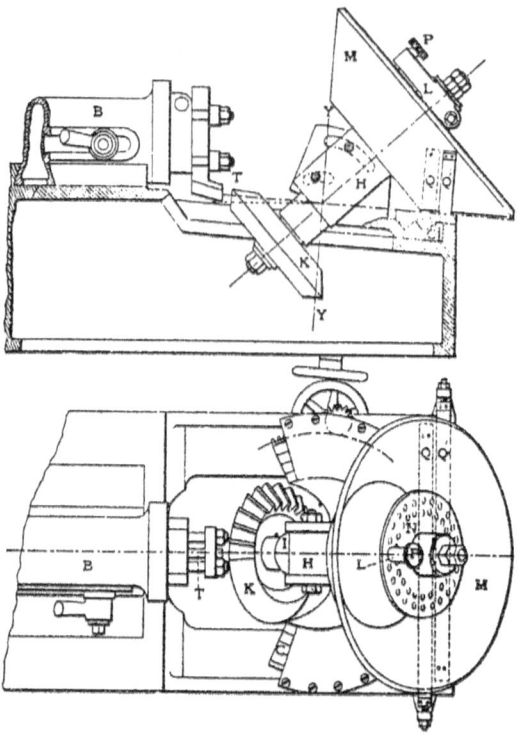

Figure 5-61 Bilgram's Bevel-Gear Generator of 1884

The Gleason mechanism was similar to figure 5-61 but unlike Bilgram employed two tools. It is a conjugate system derived from the crown gear having great circle odontoids.

The 1884, November 10th issue, of American Machinist described Oscar J. Beale's invention of a graphical method for determining the angle and diameter increments for bevel gears. Instead of using crown and master gears, only a section of the master gear's pitch cone is used, rolling on a plane representing the crown gear pitch surface. Slipping of the two surfaces is prevented by steel tapes. A straight edged tool, representing the side of a rack tooth, is used. The master gear and blank rotate in relationship with the tool. Beale also designed a machine in 1897, built by Brown and Sharpe, for finishing and correcting bevel gears by impression. A hardened crown gear ran in mesh with the bevel gear being formed.

Beale's skew bevel gears are the same as Olivier's in general theory, but his modifications and application made the gears a distinct invention.

Figure 5-63 Principle of Molding-Generating Bevel Gears

Patent #512,189 was issued in 1889 to Grant for a molding-generator bevel gear machine, the *Planoid Bevel Gear Generator*.

Automatic Gear Machines: Throughout the 18th and 19th century the technology for automatic control remained mechanical both in theory and practice. After Meikle's automatic efficiency improving turning gear in 1750 and Watt's governor the next stages were probably James Clerk Maxwell's 1868 paper on the mathematics of governors, followed fifty years later by Nikolai Minorsky's work.

During and after the Civil War the demand for machine cut metal gears was impeded by a shortage of skilled machinists creating a need for automatic machine tools. Gage, Warner and Whitney of Nashua, New Hampshire built the first fully automatic gear-cutting machine since Polhelm (1729) in 1860. It could only cut spur gears, but to eight feet in diameter and used a formed milling type cutter. They used the principle of the weighted tumbling bob to throw the clutch and reverse the cutter feed. When the cutter returned to its starting position the tumbling bob was again thrown and instantaneously the gear blank was indexed to the next tooth position. Gear cutting with a formed milling cutter only became practical with Brown and Sharpe's patented 1864 cutter.

At the 1867 Paris Exhibition Sellers displayed a gear cutting machine of the same type with the sequences controlled by stops in such a manner that the cutter could not engage the blank until it was correctly indexed for the next tooth. The machine was also designed to stop automatically when all the teeth had been cut. By use of stops it had several built in safety features. In 1876 at the Vienna Exhibition Seller's "Wheel Cutting Machine" produced teeth in which the clearance "…is only 1/100th of the pitch."

Brown and Sharpe started the manufacture of automatic gear cutting machines in 1874. They introduced two designs by Edward H. Parks, one larger machine for general purposes and the other machine for manufacturing bevel and spur gears. By 1877 three companies, Gould and Eberhardt, Brown and Sharpe in the U. S. and Craven Brothers in England would have similar automatic gear production machines for sale. From this time on there would be a proliferation of automatic machines for the production of small to medium sized spur and bevel gears.

The Gleason Works developed an automatic bevel gear generator using rotary cutters. The rotary cutter finished both sides of the tooth space in a single traverse across the blank's face. U.S. patent #605,249. (1898).

Cutters: The first use of a worm sectioned to provide cutting edges was by the English instrument maker Jesse Ramsden in 1768. Bodmer (1786 -1864) in Manchester, England, used a steel worm to hob worm wheels, and patented the method No. 8070 in 1839. The actual cutters are on loan to the London Science Museum, Kensington. The patent included an internal gear cutting machine which could also cut spur, bevel, worm gears and racks. For cutting metallic gears he used a formed milling type cutter, segmented for easier heat treatment but it was very difficult to sharpen. Bodmer provided a Pentograph for truing, cutting and shaping cutters. His cutter-grinding attachment shaped the cutter by using a copper form with oil and emery.

Whitworth used a similar cutter design as a production tool to hob helical gears in 1835. Grant suggested this machine should be used when the quantity of work warrants its use. When a worm gear is to be hobbed is fixed upon the same spindle with a master worm. The hob is forced into the blank while both revolve at the proper speeds. (British Patent 6,850) Grant wrote on "Hobbed or Concave Worm Gear" tooling. "If a spiral gear is made of steel, provided with cutting edges by making slots across its teeth, and hardened, it will be a practical cutting tool called a spiral milling tool or hob...a spiral milling cutter having a great spiral angle, and therefore called a worm."

In the U.S. Stephen A. Morse was a participant in an 1878 patent for a machine tool that could produce a globoidal worm thread. The worm may be formed into a cutter to finish the wheel. The outline of the pitch surface of the worm is an arc of the wheel's pitch circle. It was alleged at the time, later disproved, that the advantage lay in the fact that the whole side of every thread in the meridian plane is always in contact with the adjacent tooth. Clerm and Morse, Philadelphia, Pa. built a machine for cutting globoidal worms in 1882, the worm was formed into a cutter and used to finish the wheel.

Beale's article, discussing the tooth arrangement of worm gear hobs was printed in the American Machinist: "In the works of Brown and Sharpe Manufacturing Company a pair of gears was wanted of the spiral or screw type, and it was thought better to make the large gear, or member, as a worm and the small member as a worm-wheel." The wheel is the driver. The worm gear has more than six times as many threads as the worm-wheel with a pitch diameter four times that of the wheel. The original hob failed as the tops of the wheel teeth were not sufficiently trimmed to clear the backs of the hob teeth which jammed. The experiments proved that accuracy was the critical factor in obtaining good results. This was achieved by laying out on paper the teeth at three or four times their size, reduced to actual size by photography, transferred onto sheet steel, and etched in. The hob has a cast-iron body to which lugs are fastened in steps as shown in figure 5-63.

Figure 5-63 Worm Gear A and Wheel B Wheel Hob

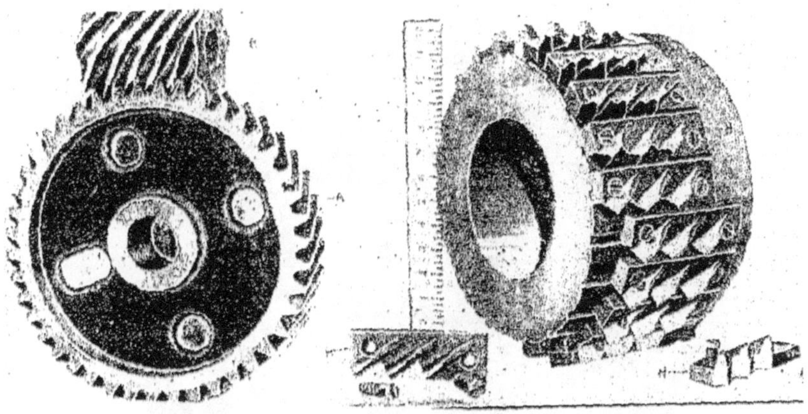

Patents were issued on November 15th 1887 and July 5th 1898 for a single tool machine acting in different positions to cut worm-wheels. The machines, and reference to the number of cutting edges required by a hob to cut a perfect worm wheel, were described in the American Machinist, May 27th 1897. In 1889 E.P. and H.C. Walker patented a machine in the U.S. that used a tool with an involute rack tooth form to cut spiral or worm gears.

Because hobbers were not generally available until the last decade of the century, the majority of worm gears were produced on lathes or milling machines. The standard tool had its sides inclined at 30° with the length and width determined by the pitch. In Dr. Weisbach's book on engineering mechanics he considers worm gears as two screws that mesh together. He writes: "If we suppose the endless screw to have a number of threads, it will assume the form of a wheel with as many teeth as the screw has threads. For example, if the screw has as many threads as the worm wheel has teeth, and if the radius r of the screw is equal to the radius R of the wheel, there will be no difference of form between the screw or worm and the worm wheel, and either can be exchanged for the other." Weisbach also concludes that in gears set at a right angle there is no friction along the threads because the teeth do not slide along the helix. Tooth friction is present due to the radial sliding of the teeth. The teeth that he depicts are of a crude square shape as shown in figure 5-64. Weisbach does suggest other forms including the involute but the gears depicted were not produced with hobs.

Figure 5-64 Weisbach Sketch Illustrating Crudity 1890 Worm Gearing

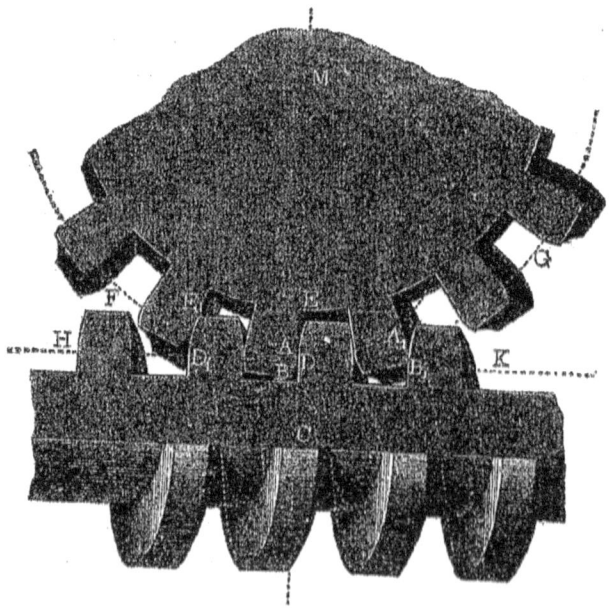

Pratt and Whitney, Hartford, Conn. built two machines so that epicycloidal cutters could be duplicated. This duplication has been made a "perfect certainty" by two machines made by Pratt and Whitney (Fig. 5-66). The system adopted by Pratt and Whitney used a tooth height

of 2⅛ times diametral pitch, a blank diameter = pcd + (2 x diametral pitch). The accuracy of the handmade cutters was frequently checked by an odontoscope.

Figure 5-66 Pratt and Whitney Tool to Produce an Exact Epicycloidal Templet

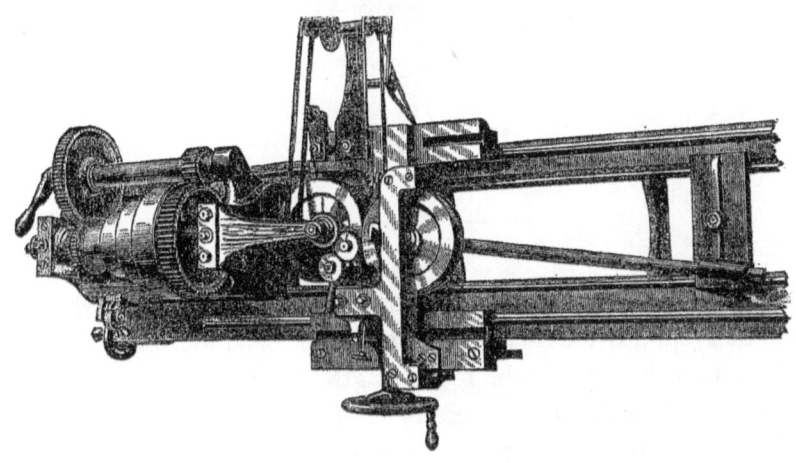

The epicycloidal milling engine invented by Ambrose Swasey used the first part in the second machine, figure 5-66, as a guide. In the second machine the contours of the gear cutter are finished by another milling operation. Until the 20th century involute cutters were supplied in sets of eight. Similar systems were designed by Brown and Sharpe. Mac Cord credits Brown and Sharpe with improving the contour of milling gear cutters with their "Epicyloidal Engine" by which the curves are automatically traced with perfect precision upon the template by continuous motion. The gear machine used an involute form relieved cutter invented by Brown in 1854 and used previously with their universal milling machine. The most important development was J.R. Brown's s design of the first true universal milling machine in 1862 (Fig. 5-46) and his invention of the formed gear cutter in 1864. The formed milling cutter accurately retained its cutting edge after successive sharpening, a necessary requirement for interchangeable gears. Toothed cutters cutting to a substantial depth had previously been in use for over twelve years but their comparatively small teeth prevented the maintenance of their accuracy. When the teeth wore the cutter was annealed, the teeth raised by peening with a hand punch, trued in a lathe, and by use of a templet the form was filed, and rehardened.

Fig. 5-67 Swasey's Epicycloidal Milling Engine

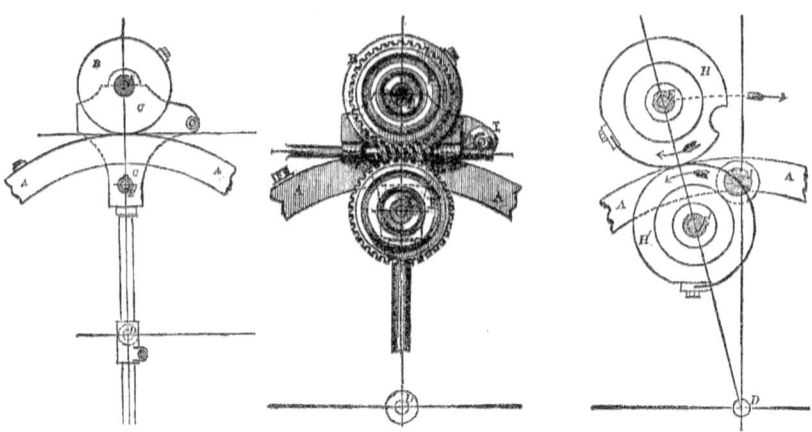

Brown's patent was for a formed cutter with segmental teeth whose face could be ground without changing its profile. Each tooth had a uniform cross section throughout the tooth length, conforming precisely to the required tooth form. The necessary cutting clearance was obtained by inclining to crown of each tooth in relation to the cutter's circumference. Sharpening merely required grinding the face of each individual tooth; this could be repeated until the teeth were too weak.

Other cutter improvements quickly followed: notched in 1869, inserted teeth in 1872 and face milling with inserted teeth in 1884. In the next fifty years Brown and Sharpe would become the major supplier of formed cutters, and the machines that used them, for both U.S. and industrial Europe.

Figure 5-68 Brown and Sharpe Method of Making a Ten Inch Hob

The 14 ½° pressure angle proposed by Professor Willis was also adopted by Brown and Sharpe, and introduced as the 14½ ° involute-tooth diametral pitch standard.

Professor Mac Cord's book titled "Kinematics" was published by John Wiley in the U.S. and by Chapman and Hall in London, the 2nd edition in 1884. He wrote on the the manufacture of gear cutters …"A master tool is first made, the contour of whose cutting edge is made to conform with the first line traced upon its template, under close scrutiny with a magnifying glass, which is the only hand-and-eye referred to. This master tool is used only to give the finishing touches to other tools, whose outlines are again compared under the glass with the template, after hardening, in order to detect any distortion which may have occurred during that operation. These latter tools, having passed this test are the ones actually used in making the cutters, and are all formed that they can be ground without changing the outline of the cutting edge. Thus the same cutter can be used for making many wheels, and the same tool for making many cutters, during the working life of which the results may be exactly alike. It should also be stated that the master tool is made with but one cutting edge, fitted to one side of the tooth-outline traced upon the template, and the second or working tool is made in two parts, both cut by one edge, in reverse directions."

"Kinematics" also provides a description of the Pantagraphic Cutter Engine invented by Francis A. Pratt. The gear cutter is turned to the approximate form, notches are cut, and the duty of the engine is to put the finishing touches to each cutting edge, and provide the correct outline. In figure 5-69 the crown gear and plastic blank are mounted in the correct relationship and rolled together, the tooth of the crown gear will form the spaces and teeth of the correct shape in the blank. In the practice of the time blanks of steel, iron, putty and clay were used.

Figure 5-69 Pantagraphic Cutter Engine

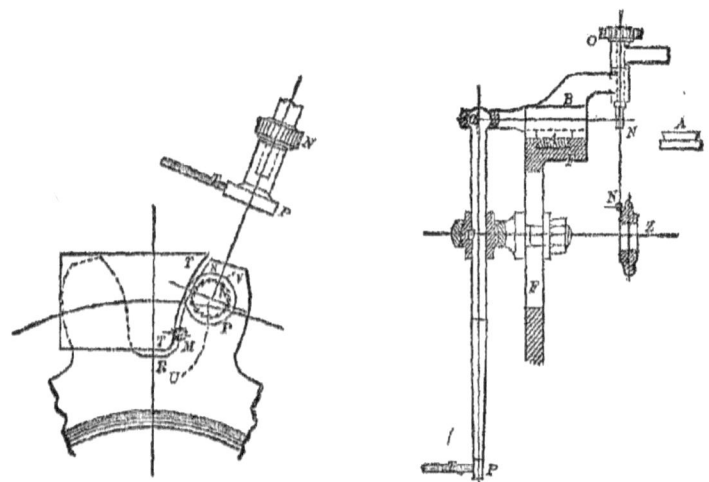

In the time before the use of automatic machinery gear cutters were shaped by a skilled draftsman who drew the tooth curve by means of rolling circles. A major advance was made with the design of odontoscopes that could check the accuracy of handmade cutters. The odontoscope method approximated the exact tooth by using circular arcs or similar easily generated curves. Grant was later in the century able to provide face and flank odontograph tables for both involute and cycloidal teeth. In order to compare the actions of various tooth systems under different conditions gear manufacturers in the 19th century made wide use of odontoscopes. The last page of Mac Cord's book "Kinematics of Machinery" was an advertisement for made to order Pratt and Whitney cutters, ref. page 232. 1880.

In England, John Holroyd and Co. Ltd. in their 1888 catalog advertised a wide range of machine tools, including a "Grinding Lathe" with covered slides and a tapering attachment. Among the large number of cutters there is a special mention of "cutters for the teeth of gear wheels which can be sharpened by grinding without changing their form". There are also "epicycloidal patent cutters" which are claimed to cut "perfectly interchangeable gears" Barber-Coleman began the manufacture of gear cutters in Rockford, Ill. in 1889. The 1880 U.S. Census illustrated inserted cutters. An example given used a solid wrought iron disc, two inches thick and weighing 400 lbs.

Figure 5-70 Inserted Tooth Cutters

The twenty-eight inch diameter disc has eighty-four teeth inserted in the rim, on edge and facing alternately, allowing a tooth every inch of circumference. The cutter was driven by a steel worm and the whole head was either moved by hand or power through a worm gear from cone-pulleys.

Gear Tooth Design: The 19th century would see enormous advances in all aspects of gear technology however; wooden gears and epicycloidal tooth forms would remain in popular usage. Almost without exception all cast gears would be of the epicycloidal form. Even though this involved separate molds for different gears. The design of gear teeth was not generally recognized as a science until the last half of the century, despite the long history of mathematical gear theory that included Nicholas's cycloidal curve in 1451, epicycloids by Dürer in 1525, De la Hire's mathematical treatise on gear design in 1694, and Camus's work in 1733. The many different gear solutions were still being obtained by trial and error. The problems were compounded when the change was made from wood to the less forgiving metal. A great deal of our knowledge of 18th century technology is derived from French authors. Known as "The Age of Enlightenment" it was the first time the intelligentsia became interested in technology. The great industrial developments, especially in metals, had occurred in Britain without the benefit of scientific observation or a written record.

Todhunter and Pearson stated in their "History of Elasticity and Strength of Materials", when referring to the technical publications of the time, "nothing shows more clearly the depth to which English mechanical knowledge has sunk." To the contrary the early 1800's brought gear theory into practical use due to the work of three Englishmen, known as the "Translators". The first of the three "Translators' was Robertson Buchanan, to be followed by, John Hawkins and Robert Willis. In his "Essay on the Teeth of Wheels", printed in Edinburgh, 1808, Buchanan put the works of Camus and de la Hire in a cogent form. He also wrote "Practical Essays on Mill Work and Other Machinery". Other English contributors were Dr. Young, James White and Sir G.B. Airy.

The Reverend William Farish was a Cambridge University student and when he developed as a brilliant mathematician with scientific interests he became a significant professor at Cambridge. Farish published four volumes; Metals and Minerals, Animal and Vegetable Substances, On the Construction of Machines, and a general discourse on hydraulics and civil engineering. The latter two volumes were published in 1813 and 1821 and are considered to be the beginning of engineering education at Cambridge University. During his lectures he exhibited models that were made mostly of metal, and similar to a meccano set, could be rebuilt in several different configurations. Models were made of the more important machines of his time and. used as an aid to his lectures. Figure 5-71 is of particular interest as it was used to demonstrate the principles of power transmission.

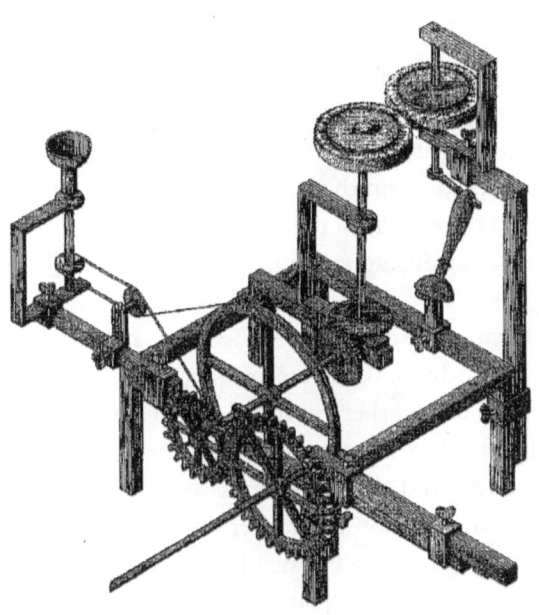

Figure 5-71 Farish's Model to Demonstrate Power Transmission Principles

Charles Etienne Louis Camus, a French mathematician, clarified the modern principles of gearing in 1733. He was a child prodigy, who lectured on mathematics in Paris at the age of twelve, and later published a two volume "Course of Mathematics." In the second volume he gave the fullest and clearest treatment on the subject of gear teeth available to that time. The first English translation appeared in London in 1806. His study covered the fully developed epicycloidal gear theory of the day including lantern, pinion/gear, spur, crown and bevels. Camus made further progress on the use of the epicycloid curve, recognizing sliding as the principal cause of friction and wear, but made no mention of the involute. Camus used the rolling cone principle to analyze bevel gears. He also considered the proper form for the ends of the teeth and the minimum number of teeth that should be used.

The Englishman, John Isaac Hawkins, attended college in Jersey, Pennsylvania and was a patent agent and consulting engineer in London. In the 1830's, Hawkins made a detailed inquiry into gearing practice. His survey included the companies of Bramah, Clement and Sharp, Maudslay and Field, Rennie Brothers, Roberts and Co. and leading clock and instrument makers. He was known as the second Translator and wrote on the results of his inquiry into English gear practice. "... there is a lamentable deficiency of the knowledge of principles, and of correct practice, in a majority of those respectable houses in forming the teeth ...". He also supplied basic rules for gear manufacture. In his interviews one manufacturer said "We have no method but the rule of thumb", clearly the gear makers relied for the most part on their own devices. Others set one leg of a pair of compasses on the pitch circle, and the other leg described a circle segment for the offside of the next tooth. In other shops the compass point would be set at different distances from the tooth center, nearer or further.

Hawkins translation of Camus's book was titled "Teeth of Wheels", and was published in 1838. He wrote an appendix in the edition: "Since M. Camus has treated of no other curve than the epicycloid, it would appear that he considered it to supersede all others for the figure of the teeth of wheels and pinions. And the editor must candidly acknowledge that he entertained the same opinion until after the greater part of the foregoing sheets were printed off; but on critically examining the properties of the involute with a view to better explaining of its application to the formation of the teeth of wheels and pinions, the editor has discovered advantages which had before escaped his notice, owing, perhaps, to his prejudice in favor of the epicycloid, from having, during a long life, heard it extolled above all other curves; a prejudice strengthened too by the supremacy given to it by De la Hire, Doctor Robison, Sir David Brewster, Dr. Thomas Young, Mr. Thomas Reid, Mr. Buchanan, and many others, who have indeed, described the involute as a curve by which equable motion 'might' be communicated from wheel to wheel, but scarce any of whom have held it up as equally eligible with the epicycloid; and owing also to his perfect conviction, resulting from strict research, that a wheel and pinion, or two wheels, accurately formed according to the epcycloidal curve, would work with the least possible degree of friction, and with the greatest durability.

But the editor has not sufficiently adverted to the case where one wheel or pinion drives, at the same time, two or more wheels or pinions of different diameters, for which purpose the epicycloid is not perfectly applicable, because the form of the tooth of the driving wheel cannot be generated by a circle equal to the' radius' of more than one of the driven wheels or pinions. In considering this case, he found that the involute satisfies all the conditions of perfect figure, for wheels of any sizes, to work smoothly in wheels of any other sizes; although, perhaps, not equal to the epicycloid for pinions of few leaves."

In an 1842 edition of Camus's work, also edited by John Hawkins, a series of experiments with wooden models was described to demonstrate actual thrust with different pressure angles "…the teeth of wheels in which the tangent of the surface in contact makes a less angle than 20 degrees with the line of centers, possess no tendency to cause a separation of their axes: Consequently, there can be no strain thrown upon the bearings by such an obliquity of tooth."

Table of Cutters for Teeth of Gear Wheels,

MADE BY

THE PRATT & WHITNEY COMPANY,

HARTFORD, CONN., U. S. A.

All Gears of the same pitch cut with our Cutters are perfectly interchangeable.

Diameter of Cutters.	Diametral Pitch.	Price of Cutters.	Size of Hole in Cutters.	SET OF 24 CUTTERS. For each pitch coarser than 10.	
5 inches.	$1\frac{1}{2}$	$25 00	$1\frac{1}{4}$ inches.	No. 1 cuts	12 T
$4\frac{1}{2}$ "	2	20 00	" "	No. 2 "	13
4 "	$2\frac{1}{2}$	18 00	" "	No. 3 "	14
$3\frac{3}{4}$ "	3	15 00	" "	No. 4 "	15
$3\frac{1}{2}$ "	$3\frac{1}{2}$	12 00	1 "	No. 5 "	16
$3\frac{1}{4}$ "	4	9 00	" "	No. 6 "	17
$3\frac{1}{8}$ "	5	7 00	" "	No. 7 "	18
3 "	6	6 00	" "	No. 8 "	19
$2\frac{7}{8}$ "	7	5 00	" "	No. 9 "	20
$2\frac{3}{4}$ "	8	4 50	$\frac{7}{8}$ "	No. 10 "	21 to 22
$2\frac{5}{8}$ "	9	4 00	" "	No. 11 "	23 " 24
$2\frac{1}{2}$ "	10	3 50	" "	No. 12 "	25 " 26
$2\frac{3}{8}$ "	12	3 50	" "	No. 13 "	27 " 29
$2\frac{1}{4}$ "	14	3 50	" "	No. 14 "	30 " 33
$2\frac{1}{8}$ "	16	3 00	" "	No. 15 "	34 " 37
2 "	18	3 00	" "	No. 16 "	38 " 42
$1\frac{7}{8}$ "	20	3 00	" "	No. 17 "	43 " 49
$1\frac{13}{16}$ "	22	3 00	" "	No. 18 "	50 " 59
$1\frac{3}{4}$ "	24	3 00	" "	No. 19 "	60 " 75
$1\frac{3}{4}$ "	26	3 00	" "	No. 20 "	76 " 99
$1\frac{3}{4}$ "	28	3 00	" "	No. 21 "	100 " 149
$1\frac{3}{4}$ "	30	3 00	" "	No. 22 "	150 " 299
$1\frac{3}{4}$ "	32	3 00	" "	No. 23 "	300 Rack.
				No. 24 "	Rack.

The cutters are made for diametral pitches. By diametral pitch is meant the number of teeth per inch in the diameter of the gear at pitch line. Two pitches should always be added to this diameter in preparing a gear for cutting. For example : a gear of 80 teeth, 8 to the inch, diametral pitch, would be 10 inches on pitch circle, but the gear should be turned $10\frac{2}{8}$ (or $\frac{1}{4}$). The teeth should always be cut two pitches deep beside clearance.

The cutters are made for a clearance of $\frac{1}{16}$ of the depth of the tooth ; example : 8 to the inch has a clearance of $\frac{1}{64}$; therefore the tooth should be cut two pitches ($\frac{1}{4}$) and $\frac{1}{64}$ deep. The gears must be set to run with this clearance to give the best results.

In ordering bevel gear cutters, give the diameter of gear at outside pitch line, and number of teeth, also the width of face. For the present all cutters are made to order

With Joseph Clement, the superintendent and chief draftsman of the Bramah Works, Hawkins experimented to determine the relative end-thrust of both forms and reached the conclusion that the advantage lay with the involute form. The oblique action of the teeth resulted in high thrust on the inadequate bearings of the time resulting in preferential use of the cycloidal tooth until near the end of the 19th century. Clement was an exceptional draftsman, an inventor, and had a leading part in the development of the screw-cutting lathe and the planer. He also introduced the tap with a small squared shank to clear the hole. Clement is also the first in Britain to use revolving cutters.

In the early part of the century several books related to gear engineering provided an insight into the then current technical knowledge. Dr. Thomas Young in his "Lectures on Natural Philosophy and the Mechanical Arts", London, England, 1807, deduced: "If the face of the teeth, where they are in contact, is too much inclined to the radius their mutual friction is not much affected, but a great pressure on their axes is produced and this occasions a strain on the machinery, as well as an increase of friction on the axes."

The book *"Century of Inventions"*, written in England by James White (1821), described helical, double helical, and herringbone designs. G. & C. and H. Carvill in New York City published a booklet on gear technology in 1832 written by Sereno Newton, and revealing knowledge of recess action. Three important books were written by Jean Victor Poncelet, the French military engineer and Professor of Engineering Mechanics in Metz (1825-35 and Paris (1838-48). His 1822 book *"Traité des propriétés projectives des figures"* details the development of projective geometry and introduced the use of mathematical and physical principles in machine design. He knew that the foci of a conic can be considered at the intersection of the tangents at the conic through these circular points. Poncelet developed an approximate method for drawing tooth profiles that was known in Germany as the *general method (allegemeine Verzahnung)*. *"Cours de écanique appliqué aux machines"* in 1827 advanced the use of mathematics and the physical principles of machine design. Poncelet's third book, of importance to gear development, was *"Introduction a la mécanique industrielle"* issued in 1829. He detailed the advantages of the involute tooth form and helped to make it the form of choice. Poncelet also introduced tensile-test diagrams to compare the properties of various irons and steels.

Pin Tooth and Wood Gearing: When they were in common use it was believed that the pin tooth gear was the best selection especially for high speeds. The wooden teeth are given only to one member, usually the driver, and a second gear was held for replacement. If the teeth were on the wheel they were expected to last longer because of the fewer contacts. Radial flanks were recommended so that the fiber of the wood may be made parallel to the flanks. Although the pin tooth is one of the oldest of gears its complicated theory and resulting defects was not properly understood until the publication of Mac Cord's book "Kinematics".

Figure 5-72 Wood Teeth or Cogs

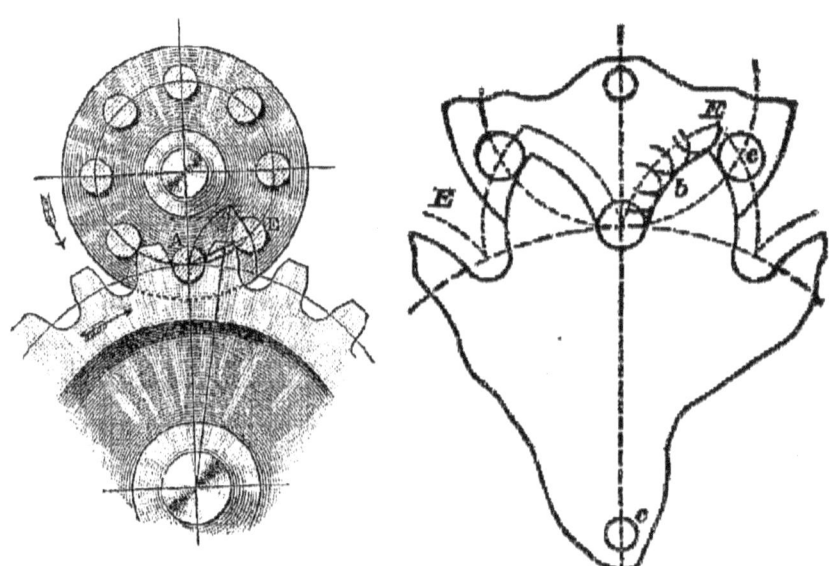

Ellis A. Davidson's book "Drawing for Machinists and Engineers" was printed in London in 1870. A typical drawing from the book to illustrate wheel construction is shown in figure 5-71. The following extract from his book illustrates the then current state of the gear art: "Cog-Wheel...a name generally understood to mean a wheel in which the teeth are made of wood and mortised separately into an iron rim, in contra-distinction to spur wheels, in which the teeth are of iron and form part of the rim itself. These wheels were each other less than if both in general use in large machinery before the introduction of cast-iron wheels and are still very common in mills. The cogs are kept in their place by pins placed inside the rims, or by keys placed between their ends. In regard to such wheels

Professor Willis says: The above construction of a toothed wheel had been partially imitated in modern mill-work, for it is found that if in a pair of wheels the teeth of one be of cast-iron and the other of wood, that the pair work together with much less vibration and consequent noise, and that the teeth wear each other less than if both of the pair had cast-iron teeth...in the best modern engines one wheel of every large-sized pair has wooden cogs fitted in the manner just described..."

Figure 5-7 Wood Teeth or Cog

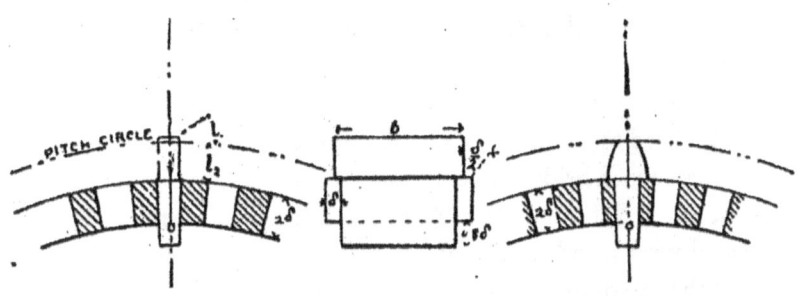

Figure 5-73 Six Arm Forty-Eight Toothed Gear Davidson Drawing

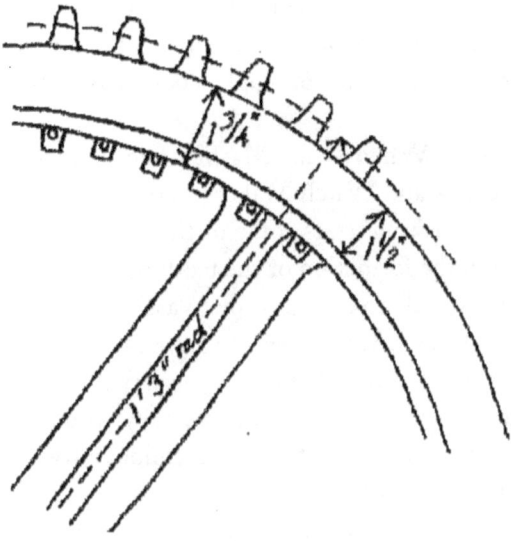

Gears operated with wheels consisting of rollers or balls also became popular. Colliers ball-worm gear and his drive with four worms and

thirteen balls were prime examples of this technique. Even today gear systems are being offered with rollers replacing the teeth. Examples of the 19th century gears are shown in figure 5-72.

In his 1889 text book "Elements of Machine Design" it is perhaps surprising to read that Professor Klein is still advocating the use of wooden gears. He writes that wooden teeth are better adapted to high speed because their shock absorption is superior to iron teeth. Klein provides arguments for using the wooden teeth in the larger gear, or if there is a spare, as a pinion in the smaller gear. This gear system became virtually obsolete with the introduction of machine cut gearing. Prior to that time they were widely used, preferably with maple gear teeth, for the majority of gear drives.

With the expansion of railroads iron had largely replaced wood in bridges. This led to a number of fatigue failures that were not understood. The failures caused bridges to collapse without warning under the train's load. The New York and Erie Railroad ordered all metal structures to be replaced with wood shortly after 1850. These misunderstood failures restricted the use of iron for gears for some time.

The Pittsburgh-Duquesne incline railway was built in 1847, and converted from steam to electric power in 1930. It was updated with a maintenance free system in the 1970's. The original gear drive is still in use and replacement aged rock maple teeth are kept readily available. Dudley reported on a 1967 visit to a New Hampshire company that they were still supplying "Tens of Thousands of maple wood gear teeth."

Clock Gears: When the Mathematical Instrument Makers, Chronometer, Clock and Watch Makers were questioned by Hawkins on their methods for making gears some of the answers were: "We have no rule but the eye in the formation of the teeth of our wheels." Others, "We draw the tooth correctly to a large scale to assist the eye in judging of the figure of the small teeth." Another, "In Lancashire, they make the teeth of watch wheels of what is called the bay leaf pattern; hey are formed all together by the eye of the workman; and they would stare at you for a simpleton to hear you talk about the epicycloidal curve." The instrument makers thought the bay leaf pattern was too pointed and it was modified by rounding the tooth tip. Connecticut was the location for the start of the U.S., clock industry and today still manufactures sixty percent of U.S. clocks.

Cycloid: The epicycloids were discovered by Albrecht Dürer in 1525. Cycloidal curves were not used because they possessed some special value, but until the 20th century, they were the simplest tooth form to make and, provided a strong enough tooth with a constant velocity/ratio. Previously alternate tooth forms had been neglected "under the impression that it was not only a very complex affair, but also it had no practical value, and as a result teeth have been made whose forms are only rough approximations to any accurate shape, and the working of the gear has been intolerably noisy and disagreeable." The Cycloid curve is traced by a point in the circumference of a circle which rolls upon its tangent. Cycloidal teeth have two distinct curves one above and one below the pitch line. The cycloid, epicycloid and hypocycloid are developed into the cycloid gear tooth form. The cycloidal form remained predominant through most of the 19th century and it was largely due to the work of Grant that the involute rightly gained ascendancy. This preference was largely due to the unfounded belief that the pressure on the shafts from the oblique tooth angle was increased with involute teeth. The gear manufacturers often misunderstood the epicycloid theory and built gears by rule of thumb. About mid century the mathematicians started to be consulted, unfortunately for the industry these first consultants dealt only in epicycloid theory.

Figure 5-75 Cycloidal Tooth Form

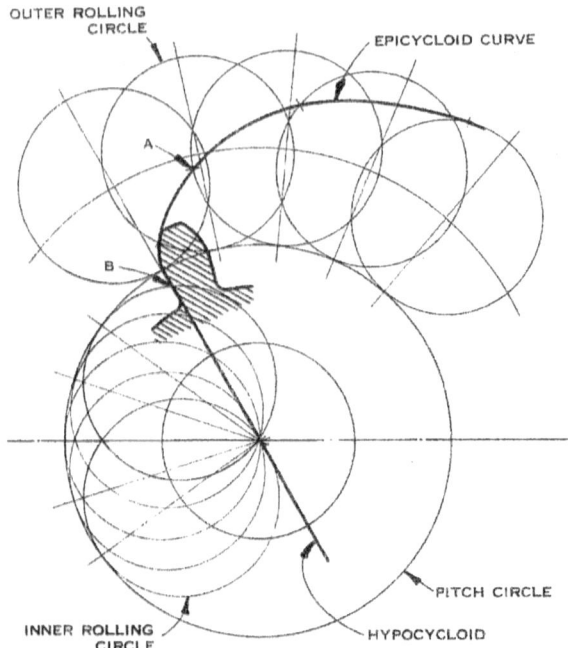

The popular cycloidal tooth forms included the Epi- and Hypo-Cycloidal versions. When a circular disc rolls along the side of a straightedge and a pencil point is on the disc rim so that it will trace a curve, that curve is a cycloid. When the disc rolls on the edge of another circular disc the curve traced is an epicycloid, and if rolled on the inside of a ring it would be a hypocycloid. Geometricians continued to use circles as profiles, and for two hundred odd years preferred the cycloid tooth form to the involute.

Mac Cord describes methods for drawing the various curves which he defines as: *Epicycloid* traced by a point in the circumference of a circle rolling inside another. *involute* is in a manner the converse of the cycloid, being generated by the rolling of a tangent right line upon a circle. It may also be seen to be generated by unwinding an inextensible fine thread from a cylinder: the thread being always taut and always tangent to the circle. *Epitrochoid* is used in a general sense, including all rolled curves, in a special sense, the curve traced by rolling of one circle upon another, when the marking point is not situated upon the circumference.

Figure 5-76 Development of cycloid, epicycloid and hypocycloid.

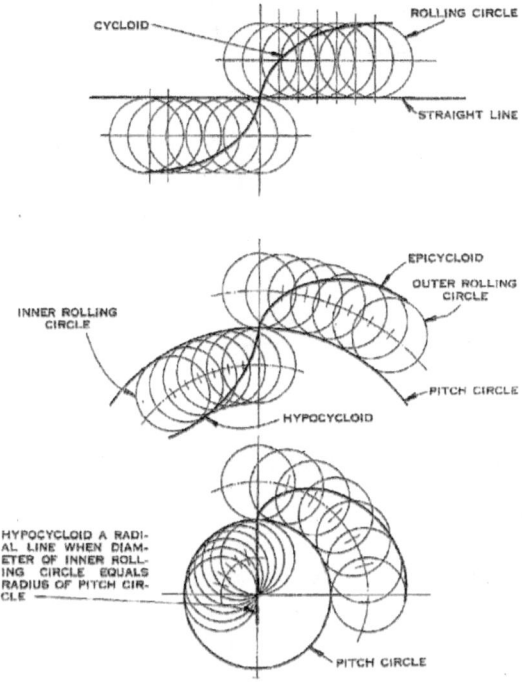

In Germany, Professor Franz Reuleaux, Director of the Royal Industrial Academy, Berlin, wrote "Kinematics of Machinery, Outlines of a Theory of Machines" in 1876. This was an important work providing the concepts of modern machines. "In earlier times men considered every machine as a separate whole…missed entirely, or saw but seldom the separate group of parts which we call mechanisms." Reuleaux classified several curves under the general name Trochoid as shown in Chart 5-2: A curve produced by any point on the radius, or extended radius, of a circle rolling along a straight line. He was recognized in his own time as the leading authority on mechanical engineering subjects associated with machine design. Reuleaux wrote in his book on machine kinematics: "I believe that within a few decades it will be the rule to employ spur-wheels working without any clearance."

Chart 5-1. Reuleaux Curve Classifications

DESCRIBING POINT.	TROCHOID.				
	EXTERNAL CONTACT.			INTERNAL CONTACT.	
	CIRCLE ROLLING UPON STRAIGHT LINE (LINEAR TROCHOIDS.)	CIRCLE ROLLING UPON CIRCLE. (EPI-TROCHOIDS.)	SMALLER CIRCLE, ROLLING. (HYPO-TROCHOIDS.)	GREATER CIRCLE, ROLLING (PERI-TROCHOIDS.)
On circle	Cycloid.	Epicycloid.	Hypocycloid.	Pericycloid.
Beyond circle	Curtate trochoid	Curtate epitrochoid.	Curtate hypotrochoid.	Curtate peritrochoid.
Within circle	Prolate trochoid.	Prolate epitrochoid.	Prolate hypotrochoid.	Prolate peritrochoid.

Sir George Biddell Airy, Lucasian professor of mathematics, Cambridge, wrote "On the Forms of the Teeth of Wheels" in 1825. The current epicyclic problem was that given the form of one wheel what form would be required for the other for them to work together. Airy gave the mathematical solution and proof that applied to any tooth.

Professor Robert Willis introduced the "Buttressed" Tooth in an Institute of Civil Engineers, London, 1838 paper. The design was to increase the tooth strength of gears continuously operating in one direction. In his words it was accomplished by increasing the angle of obliquity of the back of the tooth. When done correctly the gears could operate in either direction but there is increased pressure on the bearings when using the back face of the teeth due to the increased obliquity of action. The limit he placed on pressure angle was 32°. In 1841 Willis wrote the first English textbook on "Principles of Mechanism" in which he included under the section "Sliding Contact" a velocity ratio constant, and tables for rolling contact and tooth forms. In the book he quoted Camus's law "If the pinion is to turn the wheel with a uniform force, the curve of its leaf, and that of the tooth of the wheel must be generated in the manner of epicycloids by one and the same describing circle, which must be rolled within the circle of the pinion to describe the inner form of the leaf, and on the outside of the circle of the wheel to describe the outer form of the tooth," etc. One third of the four hundred pages is devoted to gearing. The book was the first to provide the correct basis for the interchangeability of cycloid gearing: "If for a set of wheels of the same pitch a constant describing circle be taken and employed to trace those portions of the teeth which project beyond each pitch line by rolling on the exterior circumference, then any two wheels of this set will work correctly together… the diameter of the constant describing circle shall be made equal to the leas radius of the set." In addition to tooth forms Willis describes epicyclics and differentials.

In order to overcome the imperfect spur gears that were being used, in 1858 W.N. Nicholson, patented in England, the use of internal gears to drive rolling mills. In order to eliminate noise a later modification was to fill the hollowed out wheels with shot. Hawkins was the first to appreciate the value of the involute curve, and its advantages over the epicycloids, and the proper basis for cycloidal gear interchangeability. He wrote that short teeth provided a stronger tooth because the tooth base did not have to be relieved. Radius, not diameter, should be the basis for generating epicycloidal teeth. Hawkins describes a method for correcting the tooth form after they have been roughly formed, a method devised by Joseph Saxton. (p.206)

The January 1877 issue of the Franklin Institute published A.K. Mansfield's law of internal cycloidal interference. The following conclusions were prevalent on gears with a cycloidal tooth form.

1. Distance apart must be exactly equal to the sum of the radii of the pitch circles.
2. Considerable backlash must be present to allow for inaccuracies as unlike the involute the centers cannot be adjusted to improve the backlash.
3. The tooth apex has a higher tendency for breakage.
4. The form is more difficult to construct.

Thomas Box, Professor of Mechanical Drawing and Design at Stevens Institute of Technology, Hoboken, N.J., wrote a book, based on his practical experiences, that was of great historical significance. "Mill-Gearing" was published in London and New York in 1869. In the introduction Box states that the book was written to try and present a standard design method that would replace the unsatisfactory methods currently used. Useful charts such as chart 5-2 and drawings of wheel and tooth design figure 5-75 were provided.

In the 4th edition dated 1888, on the subject of epicycloidal teeth Box wrote: "… epicycloidal and other forms of teeth do not possess, namely, that the distance of the centers of a pair of wheels may be varied so that the teeth are more or less deeply in gear without affecting the regularity of the motion, ...With epicycloidal teeth the distance of centers must be strictly preserved, and where this can be done, which is the case in most instances, that form of teeth is the best."

Chart 5-2 Tooth Proportions

TABLE 3.—Of the PROPORTIONS of TEETH, &c., &c., for SPUR-WHEELS.

Pitch in Inches.	Constant for the Diameter.	IRON-AND-IRON-TEETH.					MORTICE-WHEELS.						
		Length of Teeth.			Thickness of Teeth.	Clearance.	Length of Teeth.			Thickness of Teeth.		Metal at end of Mortice.	Depth of Rim.
		Above pitch.	Below pitch.	Total.			Above pitch.	Below pitch.	Total.	Wood.	Iron.		
1	·3183	·344	·469	·813	·45	·100	·250	·375	·625	·60	·4	·500	1¼
1¼	·3979	·430	·570	1·000	·57	·112	·312	·452	·764	·75	·5	·592	1½
1½	·4775	·516	·669	1·185	·689	·123	·375	·528	·903	·90	·6	·681	1¾
1¾	·5570	·602	·767	1·369	·809	·132	·438	·603	1.041	1·05	·7	·768	2
2	·6366	·688	·865	1·553	·930	·141	·500	·677	1·177	1·20	·8	·853	2¼
2¼	·7162	·774	·951	1·050	·150		·563	·751	1·314	1·35	·9	·938	2½
2½	·7958	·860	1·058	1·918	1·171	·158	·625	·823	1·448	1·50	1·0	1·020	2¾
2¾	·8753	·946	1·153	2·099	1·292	·166	·688	·895	1·583	1·65	1·1	1·102	3
3	·9549	1·032	1·248	2·280	1·414	·173	·750	·966	1·716	1·80	1·2	1·183	3¼
3¼	1·0345	1·118	1·343	2·461	1·535	·180	·813	1·038	1·851	1·95	1·3	1·263	3¼
3½	1·1141	1·204	1·438	2·642	1·657	·187	·875	1·109	1·984	2·10	1·4	1·343	3¾
3¾	1·1936	1·290	1·532	2·822	1·778	·194	·938	1·180	2·118	2·25	1·5	1·421	4
4	1·2732	1·375	1·625	3·000	1·900	·200	1·000	1·250	2·250	2·40	1·6	1·500	4¼
(1)	(2)	(3)	(4)	(5)	(6)	(7)	(8)	(9)	(10)	(11)	(12)	(13)	(14)

NOTE.—See also the scales in Plate 4, which give the dimensions in this Table by direct measurement.

Figure 5-77 Box's Wheel and Tooth design

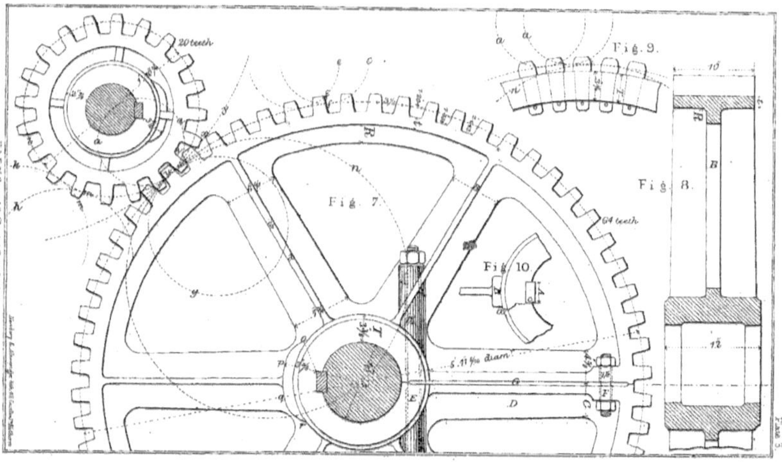

Involute: The involute history was described in Chapter Three by Robert Buchanan in his *"An Essay on Gear Teeth"* in 1808 makes reference to Professor Robinson of Edinburgh's earlier work on the involute tooth with an illustrative diagram as shown in figure 5-76. Buchanan clearly understood the involute principles and in reference to this figure he wrote: *"It is obvious that these teeth will work before and after passing the line of center, they will work with equal truth, whether pitched deep or shallow, a quality peculiar to them and of great importance."*

Figure 5-78 Buchanan's Diagram of Involute Tooth Action 1808

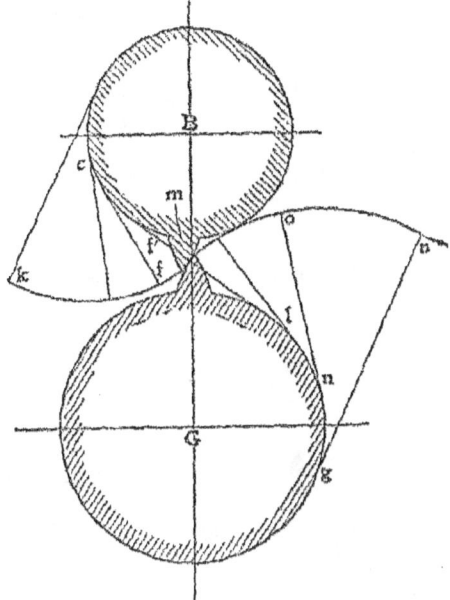

Sereno Newton's note book on gear engineering includes the statement on wheel teeth "The proper form for the curved part of the tooth of a wheel or pinion to act on a rack is an involute of a circle." According to Professor Willis the involute gear tooth was first suggested by Euler, in his second paper on "The Teeth of Wheels."

In the U.S. Grant wrote the "1885 Handbook of the Teeth of Gears". In this book he proved beyond any doubt the advantage of the involute form over the epicycloid, even though at this time the epicycloidal tooth form was the most popular. Grant also promoted gear standardization, and made many contributions to advance gear technology. He also wrote "Ordontics or the Theory and Practice of the Teeth of Gears" in 1891, followed by "Gear Book for 1893", and "Why Gear Hobbing Machines Cut Flats". All were of major assistance to the gear industry.

Bevel Gear Forms: The English engineer, Thomas Tredgold in 1822 provided a close approximation method for the drawing of bevel gears that would remain in use for decades. In 1869 Rankine also credited Tredgold with the best way for the laying out of bevel gears and, the understanding of frictional gearing to David Kirkaldy. "… to increase the friction or adhesion between a pair of wheels which is the means of transmitting force and motion from one to the other, their surfaces of contact are sometimes formed into alternate ridges and grooves parallel to the plane of rotation.… angle between the sides of each groove is about 40°." Kirkaldy also wrote "The surfaces of the teeth of a skew-bevel wheel belong, like its pitch surface, to the hyperboloidal class, and may be conceived to be generated by the motion of a straight line which in each of its successive positions, coincides with the line of contact of a tooth with the corresponding tooth of another wheel."

Figure 5-79 Development Mating Cones

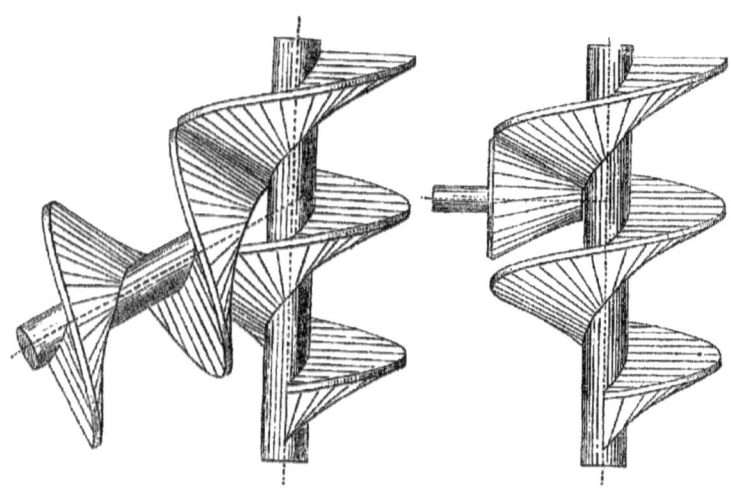

Professor Earle Buckingham in his 1949 treatise wrote "Fortunately we do not have to resort to the solution of bevel-gear problems, on the surface of a sphere because we have Tredgold's Approximation, which reduces the problem to one of spur gears, and is sufficiently accurate for all practical purposes for tooth numbers greater than 8." The tooth curves were drawn on cones, tangent to the sphere at the pitch line.

In 1823 Poinsot presented his celebrated memoir "Théorie nouvelle de la Rotation des Corps" to the French Institute containing important theorems related to gear geometry. "the rotary motion of a body about an axis which incessantly varies its position around a fixed point is identical with the motion of a certain cone whose vertex coincides with this point, and which rolls, without sliding, on the surface of a fixed cone having the same vertex." The surface of any cone can be developed upon that of any other cone. His reasoning included force and velocity with changes in motion, ideas that were revolutionary for the time.

In Paris the French mathematician Th. Olivier wrote "Théorie géométrique des engrenages" in 1832. He was the first to propose a method of generating conjugate gear teeth based on the application of the surface of a tool. He described a spiraloidal tooth on an involute skew bevel gear. See figure 5-80. At Brown and Sharpe Beale's skew bevel gears used Olivier's work as a basis making improvements in a practical form and application. They were described in the August 28th 1890 issue of the American Machinist.

Grant paid Olivier the highest compliment when in his book "A treatise on Gear Wheels" when he wrote, "Indeed, the closest possible scrutiny of Olivier's theory, without the aid of Beale's experimental work, fails to detect a flaw in it." Professor Herrmann gave a law on skew bevels claiming Olivier's tooth could not be correct. Grant's opinion was that Herrmann's law was clearly wrong, and that there was no flaw in Olivier's theory: "Of all the skew tooth surfaces that have been proposed, there is but one, the Olivier involute spiraloid, that can be proved to be theoretically correct." The pitch surface of the skew bevel gear was known as the hyperboloid of revolution.

Figure 5-80 Olivier's Involute Skew Bevel Gear

Grant wrote extensively on the subject of bevel gear tooth forms. He described the cycloidal and involute bevel forms, skew bevels, twisted skew teeth and even skew pin gearing, hypoid, and pitch hypoids: "The utility of the hypoid as the pitch surface of the skew gear depends upon the peculiar property that any number of such surfaces will roll together, and drive each other by frictional contact with velocity ratios in the proportions of their skew angles, if their gorge radii are in the proportions of the tangents of their skew axles."

Micel Chasles, who taught mathematics at the École Polytechnique and in 1846 was Professor of Geometry at the Sorbonne, was credited with the first understanding of the motion of conical forms. He wrote his conclusions in his book "des Méthodes en Géométrie" published in 1837.

In 1864 Belanger had his conclusions published that were an extension of the work of Poinsot. The publication named "Traité de Cinématique" included two major proposals and the second more importantly was to consider the position of the axis as forming a pair of ruled surfaces, one for each body, so that the motion is reduced to a rolling of the two ruled surfaces upon each other and simultaneously sliding end long upon each other of the generators in contact.

A book "507 Mechanical Movements" written by Henry T. Brown was published in 1868. Numerous gear mechanisms were illustrated and described. In addition to the commonly used gear types elliptical, square, segmental, etc. were all included. For example item 44 shown in figure 5-79 was used "to transmit great force and give a continuous bearing to the teeth." The spur gear system was used to drive screw propellers, racks, and large iron-planing machines.

Figure 5-81 Henry Brown's Gear Systems

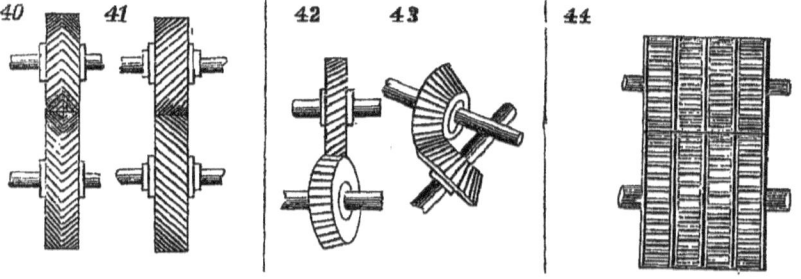

Professor Mac Cord in his 1884 book "Kinematics" included a description of his invention, the Elliptic Bevel Gear, which could operate at an inclined angle. He also gave a thorough examination of face gearing, however, in Grant's 1899 book "A Treatise on Gear Wheels" states "At the present day they (face gears) are not in use, and do not deserve much study." Mac Cord concluded the following law: "All relative motion of two bodies may be considered as the twisting or rolling of ruled surfaces or axoids." As the surfaces are always the loci of a series of axes they may be called axoids.

Figure 5-82 Mac Cord's Description of 1880 Gearing

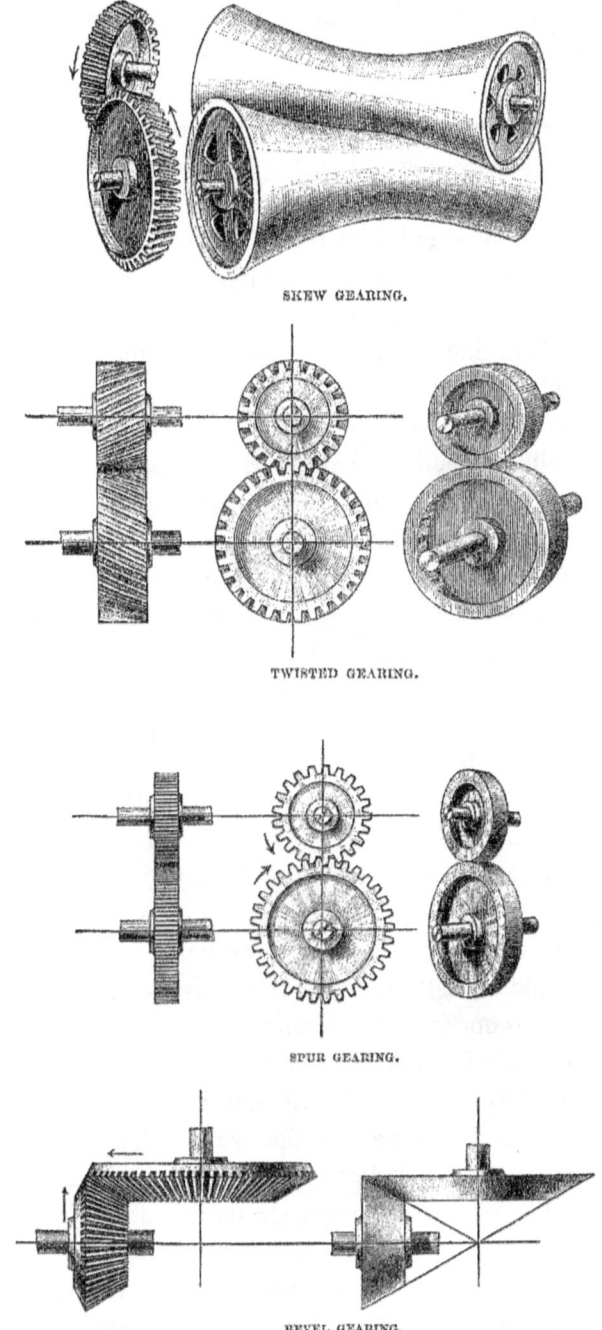

Mac Cord also wrote about the six classes of gearing (as shown in figure 5-82) that were available: spur, bevel, skew, twisted, screw, and face.

Weisbach in his 1890 "Machinery of Transmission" wrote that when the gears axes do not intersect or are parallel, the axoids represent the relative motion when a uniform velocity ratio is transmitted, can be shown to be hyperboloids. By twisting an auxiliary hyperboloid on these primary ones, surfaces can be generated that will be suitable for tooth profiles. The resulting toothed wheels are often called skew bevel gears. Profesor Herrman's section of Weisbach's book "Mechanics of Engineering and Machinery" was said by Grant to be the most important work that can be named in connection with gear teeth.

Some say Cambridge Professor Willis's most important work was the analysis of bevel gear teeth. Willis is also credited with the invention of the hypoid gear. Willis's contribution stated:" The surfaces adapted for teeth in the case of rolling hyperboloid might be obtained in a manner similar to those of rolling cones; by taking an intermediate describing hyperboloid; but it does not appear possible to derive from this any rules sufficiently simple for application." He further commented that a sufficiently close approximation could be made by drawing two cones normal to the hyperboloid frustum selected, developing them, and after laying out the teeth as in Tredgold's method, wrapping them back in their proper relative positions. (The Swedish machine builder Landop is said to have built a hypoid gear in 1741. (p.106)

In 1886 Alex B.W. Kennedy took gearing another step forward with his book "Mechanics of Machinery". On making bevel gears he wrote "It is comparatively easy to make correctly shaped patterns for the teeth of bevel wheels, and the shape of the bevel tooth, when the wheel is cast, will be fairly near the shape of the pattern. But if it is required that the profiles of the teeth should be really accurate, they must be, as with spur gearing, machined, and this operation is one providing some practical difficulties. The most recently devised machine for this purpose, a very ingenious one, is probably that of Mr Bilgram, described in Engineering, vol xl. p.21."

A.D. Pentz was the first to describe parallel depth bevel gears in an article published in the September 1891 issue of American Machinist, also it was suggested by Walter Gribben in the same issue that they be cut with an ordinary spur gear cutter.

Worm Gear Forms: In "Endless Screws or Worms and their Wheels" Dr Robinson detailed a worm form, he said "make the screw cut the teeth".

In the U. S. Hewes and Phillips Iron Works of Newark were known for their planers and use of cylindrical worm gears. Several of their worm driven planers were exhibited at the Newark Industrial Exhibition of 1873. They experimented with several pitch angles, finalizing on 20 degrees with a quadruple thread. These worms had negligible wear after twelve years of service. Their final conclusion was to use worms as small in diameter as practical with a twenty degree pitch angle, an epicycloidal tooth, case-hardened open-hearth steel for the worm and hard cast iron for the wheel.

The Willis theory of spiral tooth action was presented in 1886, it was based only on right angle worm gears, and no longer accepted the theory that the section of the two gears made by the plane act together like a rack and pinion. An article stating if there was any action at all it would be upon the normal spiral section was published in the "American Machinist" on May 19th. The angle used for worm threads was also 14½° thus making the straight sided rack of the involute system corresponding in angle and other proportions with the worm thread.

Globoidal Worm Gears: The Hindley type worm gear was considered by Henry Hindley to have superior wear qualities to the circular type. Contrary to the "encircling" theory there is line contact extending across the tooth at the pitch line. The contact is of a broader nature due to material elasticity.

Figure 5-83 Hindley Gear

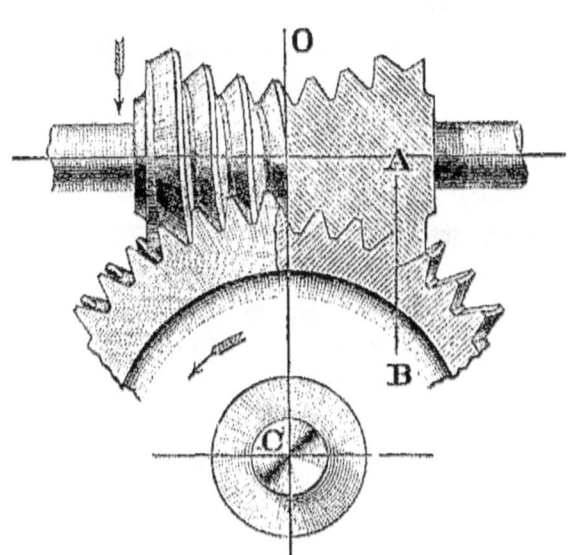

Grant wrote "...erroneously stated that the worm fits and fills its gear on the axial section...contact may be linear on some line of no great length, but it is probably point contact on the normal section." Why Grant said normal and not axial we do not know. The objections of the time were that such gears had to be set up accurately with perfect alignment and no end play. Longitudinal displacement would force the gear to cut into the other. To counter poor workmanship it was recommended that there was a large amount of backlash.

Reuleaux is credited with introducing the technically correct term globoid for the Hindley encircling gear tooth form. Another part of Reuleaux's career was in promoting and advancing German industry. He was a judge and commissioner at exhibitions from Europe to Australia. As commissioner at the Centennial Exhibition in Philadelphia he wrote his critical "Letters from Philadelphia" in which he characterized German industrial products as billig and schlecht (cheap and ugly). The letters encouraged the technical effort that led to Germany's manufacturing excellence.

In 1883 the U.S. Government started to use the globoidal tooth form for shock loads and minimum backlash. The worm had straight sides in the axis of the worm and conforms to the curvature of the wheel. Willis, Sellers, Lewis, and Thurston also studied and tested the worm gear and its forms.

Design and Circular/Diametral Pitch: J.G. Bodmer In 1843 wrote "On the Pitch of Spur and Bevel Wheels and the Shape of the Teeth of Worm Wheels and Worms Working into each other." This was the first published account of the use of circular pitch versus diametral pitch. Bodmer recognized and promoted the advantages of using diametral pitch. The system was so widely adopted in North West England that it became known as "Manchester Pitch".

Professor Unwin defined diametral pitch in 1882 as "A length which is the same fraction of the diameter as the circular pitch is of the same circumference." In his book "Elements of Machine Design" and provided for Factors of Safety. He also provided a table "Pressures on Bearings" that gave the maximum bearing value as 3,000 lb/ins^2 for slow speed and intermittent loads.

The Englishman Archibald Sharp in his 1896 book "Bicycles and Tricycles" wrote: "The American gear-wheel makers define the diametral pitch as "The number of teeth in the gear divided by the pitch diameter of the gear. The latter may be called the pitch-number."

Grant in 1899 would write "The old and clumsy circular pitch system is in universal use for worm teeth, for the reason worms are generally made in the lathe, and lathes are never provided with the proper change gears for cutting diametral pitches. The error is so firmly rooted that it is useless to attempt to dislodge it."

Many of the later writers based their work on Willis's essay "The Teeth of Wheels". Willis also issued a paper on "Circular versus Diametral Pitch". Under the heading "Synoptic Table of the Elementary Mechanisms" he listed many of the gear's basics listed as follows:

Division A – Rolling Contact – Directional Relation Constant
Class A – Velocity - ratio constant. (Rolling cylinders, cones, and hyperboloids).
General arrangements—forms of toothed wheels and Pitch.
Division B – Sliding Contact – Directional Relation Constant
Class A – Velocity -ratio constant. (Forms of the individual teeth of wheels).
Endless screws or worms and their wheels.

In 1881 Heinrich Rudolph Hertz was to provide the contact stress analysis which allowed values to be established for a variety of geometric forms. The work became the basis for bearing and gear stress and deformation calculations. His paper gained wide attention and was published in "Borchardt's Journal". In 1889 he became Professor of Physics at the Bonn University. Hertz only lived until the age of thirty-six and few have achieved so much in science in so short a time.

There were two leading Philadelphia tool builders, Bement and William Sellers and Company, the latter until 1880 made cycloidal gears with cutters of true cycloidal shape. They now changed to 14 ½° pressure angle involutes, others were using Willis's 15°, neither satisfied the requirements of an interchangeable system, and based on the recommendation of Charles H. Logue, Sellers changed to a 20° system in 1885. Willis chose 14½° for the pressure angle because the sine of the angle approximated ¼. To obtain a stronger tooth Willis recommended increasing the pressure angle within limitations. The angle also closely approximated the pressure angle used for epicycloidal teeth. Thus coinciding with Professor William John Macquorn

Rankine's book "A Manual of Machinery and Millwork" was published, in London England in 1869. About the book Grant wrote: ..."although the driest of books, its value is as great as its reputation". The book was written in three parts, Geometry of Machinery, Dynamics of Machinery, and Materials, Strength and Construction of Machinery. The book provided gear details on epicycloids, worms, bevels, materials, geometry, mathematics, rolling and sliding action, and manufacture. Rankine credited Dr. Hooke with the invention of the helical gear, and the experiments of David Kirkaldy with proving quenching in oil improved steel strength, frictional gearing was credited to Mr. Robertson. References are made to the previous works of: Willis on "Mechanisms", Fairbairn's "Millwork", Holtzapffel's "Mechanical Manipulation", Buchannan's "Millwork" edited by Tredgold and Rennie, and "Essay on Tools" by Nasmyth. The book provided the following technical data:

The smallest number of recommended pinion teeth:

1. Involute ... 25, Epicycloidal ... 12, Round teeth or staves ...6

"The least thickness:--Divide the greatest pressure to be exerted between a pair of teeth in pounds by 1,500; the square root of the quotient will be the least proper thickness in inches."

"The least breadth; -- Divide the greatest pressure to be exerted in pounds by the pitch in inches, and by 160, the quotient will be the breadth in inches.

Klein's text book "Elements of Machine Design" was produced for the engineering students at Lehigh University in 1889, and provided considerable detail on gear design. He gave credit to Reuleaux's book "Kinematics of Machinery", and credited the best method for cutting the teeth of bevel gear pinions to Professor C.B. Richards. This book detailed all aspects of power transmission.

In 1889 Oscar Lasche had demonstrated by the empirical method the advantages of the rack shift. He examined the effect it had on the life of tramway gears. Geometrical, kinematic, and dynamic studies have shown the importance of this gear element which resulted in its standardization in 1957.

Richard. Stribeck, Professor of Machine Construction, Stuttgart and Dresden, in his Dresden position he began a study of gears including experimental studies on worm gears. His experiments determined the limitations in loads and speeds of worm gears. His 1894 book "Berechnungder Zahnräder" contained a systematic analysis of gear mechanics, gear wear and its consequences. Additional work on the subject took place in 1903. Stribeck is also renowned for his work on tribology friction, especially the friction characteristics of bearings, the Stribeck Curve, and the mechanical properties of materials. He continued his studies on gear performance and by 1909 had produced twenty-three papers. Hertz, Stribeck and Goodman were the main contributors to the science upon which the rolling-element bearing industry was founded.

The British Institution of Civil Engineers published a paper in 1896 covering all the known aspects of gear technology; it was titled "Circular Wheel Teeth", vol. cxxi.

Tooth and Shaft Strength: In 1821 Tredgold made an attempt to calculate the strength of cast iron gears and had published his results in "Practical Essay on the Strength of Cast Iron." In the 1841 third edition of his "Treatise on Millwork" the Scottish engineer George Rennie added a detailed account of the strength of wood and cast iron teeth, including tables and a graph for selecting teeth of adequate strength. Box's book "Mill Gearing" until the end of the century provided the best acceptable method and amongst others would be used by Grant for estimating the strength and power rating of gearing and shafts. Thomas Box's rule was considered the most reliable method is use at that time:

$$\frac{12 \, c^2 \, f \, \sqrt{d} \, n}{1{,}000}$$

c circular pitch, f face width, d diameter, n rpm all dimensions in inches

In Chart 5-3 Box provides the anticipated horsepower of the current power sources such as the ass, donkey, horse, man, and mule, and the power rating for shafts in three materials over a wide speed range. He also emphasized the need to apply factors when selecting gears based on the type of loading, and that the load carrying strength of mortise-wheels was higher by the ratio of .05 to .043. The chart also provides us with a list of the popular gear applications of the period. The following are Box's power ratings for three popular shaft sizes based on three Materials and commonly used speeds:

Chart 5-3 Box's Power Tables

TABLE 6.—Of the POWER of SPUR-WHEELS with IRON-TEETH, from CASES in PRACTICE.

Nominal Horse-power. Actual.	Nominal Horse-power. Calculated.	Revolutions per minute.	Pitch in inches.	Diameter of wheel. ft.	Diameter of wheel. in.	Width on the Face. inches.	REMARKS.
60	51·1	19·5	3¼	8	0	9	Paper-mill; wore excessively and broke down.
60	62·4	..	..	..	..	11	Same, made wider, worked better.
50	55·9	40	2¾	15	2½	7	Sawing machinery.
46	44·8	17·5	3	12	0	8	Boulton and Watt.
45	41·1	20	3½	5	0	9	Three-throw deep-well pumps.
44	42·6	25	3¼	4	4⅝	9	Ditto Crystal Palace.
42	44·5	22	3¾	5	5	9	Paper-mill.
42	45·4	22	3¼	4	5 13/16	10	Three-throw deep-well pumps.
40	39·2	40	3	3	11⅝	8	Paper-mill.
32	30·0	19	3	8	10	6	Boulton and Watt.
30	29·6	24·5	2¾	6	9	7	Asphalte machinery.
30	35·4	22	3½	2	9	10	Sugar-mill.
*30	27·4	5·76	3¾	8	3	12	Ditto.
30	26·7	30	2½	13	1⅝	5	Sawing machinery.
25	22·1	24½	2¾	3	10	7	Paper-mill.
*22	24·3	13·6	3	5	3	7½	Three-throw pumps, Crystal Palace.
20	20·6	36	2½	4	2½	6¼	Engineering machinery.
20	18·8	15½	2½	6	10⅝	6⅝	Three-throw deep-well pumps.
20	23·0	7·28	3	6	8¼	8½	Ditto.
20	22·0	14·1	2¾	5	9½	7½	Ditto.
20	20·3	36	2½	13	4 7/16	4½	Sawing machinery.
16	19·0	60	2½	3	6½	6	Ditto. New Zealand.
16	15·8	32	2½	3	0½	6	Three-throw pumps.
12	12·1	34	2½	1	7½	6	Corn mill.
8	10·4	45	2½	1	3½	5½	Ditto.
8	8·8	24	2	3	11½	5½	Three-throw pumps.
*7	7·2	30	2	3	3½	4½	Ditto.
6	6·5	45	1¾	2	1½	4½	Ditto.
*5·5	5·5	19	2	3	4½	4½	Ditto.
4	4·5	16·4	1¾	3	11½	4½	Ditto.
4	4·2	53	1½	1	7½	4	Ditto.
1	1·1	88	1⅝	0	7	2	American horse-works.
815	822·3						

NOTE.—The wheels in this table, except those marked *, were first-motion wheels to steam-engines.

TABLE 7.—Of the POWER of SPUR-MORTICE-WHEELS, from CASES in PRACTICE.

Nominal Horse-power. Actual.	Nominal Horse-power. Calculated.	Revolutions per minute.	Pitch in inches.	Diameter of wheel. ft.	Diameter of wheel. in.	Width on the Face. inches.	REMARKS.
42	45·4	22	2¾	18	3¾	6	Centrifugal pump.
30	26·0	24·5	2¼	17	2½	5	Ditto.
30	31·7	45	2½	18	10½	5	Papier-maché works.
25	24·5	25·5	2½	14	3½	5½	Centrifugal pump.
22	18·1	26	2	14	10	4¾	Rope machinery.
16	20·0	26	2½	13	4 3/16	4½	Engineering machinery.
15	16·3	30	2	12	3⅝	4½	Agricultural ditto.
15	13·8	25½	2½	3	3 1/16	6	Three-throw pumps.
12	14·7	11	2½	6	9	6½	Ditto.
12	17·1	33	2	12	3½	4½	Crape-weaving machinery.
11	10·6	34	2	2	10½	5½	Three-throw pumps.
10	14·5	15·2	2½	5	11½	6	Ditto.
10	12·7	16	2½	5	1½	5½	Ditto.
8	9·6	16	2	5	1½	5⅝	Ditto.
8	7·0	16	2	2	11½	5	Ditto.
6	6·8	45	1½	6	1½	3¾	Envelope machinery.
4	4·6	53	1½	3	0	3½	Three-throw pumps.
4	4·4	53	1½	1	3½	3½	Ditto.
3	3·6	65	1½	0	8½	3½	Ditto.
1	1·3	100	1½	0	8½	2½	Washing machinery.
274	302·2						

NOTE.—The wheels in this table were all first-motion wheels to steam-engines. In many cases, the particulars given are not those of the wood-toothed wheel itself, but those of its fellow.

(41.) "*Power of Bevel-wheels.*"—The rules already given apply to bevel-wheels as to others, with certain modifications. Bevel-wheels, as we have stated in (26), have a maximum and a minimum diameter, and also a maximum and minimum pitch, and in calculating their power we must not use the *reputed* sizes which are always the maximum ones, but must ascertain and use the *mean* diameter and the *mean* pitch. Thus, say, we take the case of the wheel A, Fig. 12, with 30 revolutions per minute; the maximum diameter is 3 feet 7 inches, but the minimum diameter *a* is 2 feet 8 inches only, the mean is therefore 3 feet 1½ inches, or 3·125 feet. Again, the maximum pitch

Shaft diameters	2"	6"	12"
RPM	741	16	5.6
Common cast iron and wrought iron shaft			
HP	16	22	60
Common cast iron steam engine crankshaft			
HP	10.2	14	38.2
Cast iron steam engine crankshaft			
HP	6.4	8.8	24

On October 15th, 1892, the popular Lewis formula was presented to The Engineers' Club of Philadelphia by the member Wilfred Lewis in his paper: "Investigation of the Strength of Gear Teeth". He was probably the first to consider the tooth form as a factor in determining tooth strength, and that the direction of the pressure angle was always normal to the tooth profile. Lewis made the observation that it was not possible to compute the load stress on an involute gear, but it was possible to compute the stress of a load on a parabola. He assumed that at the beginning of the contact the load was concentrated at the end of the tooth. This observation was the basis for his bending strength formula. He recognized that the instantaneous tooth load was affected by the velocity of the system. He quoted John Cooper's 1879 conclusions. The formula was later revised with a speed factor ratio by Carl G. Barth based on tests and experience with cast iron gears with cast tooth forms prior to 1868.

Grant was to write in 1899: "Multiply three hundred and fifty pounds by the face of the gear, and again by circular pitch, both in inches, and the product will be the safe working load of the tooth in inches." He believed the load could be doubled if two teeth were always in contact. A hard wood mortised cog was about one third the strength of a cast iron tooth and steel had double the strength. Cut gears and cast gears had almost equal strength. In 1832 Serena Newton had calculated an allowable unit load of 75 for wood and 300 for cast iron.

High Speed: The Swedish engineer de Laval, in his 1889 development of the steam turbine advanced high speed helical gearing technology. In the June Proceedings of the Philadelphia Engineers Club in 1894 Geyellin wrote on "the highest recorded speed for gearing". The mortise bevels had a peripheral velocity of 3,900 fpm.

In 1899, the chief engineer of the General Electric Company, Oscar Lasche, Berlin, Germany, wrote an article on high speed gearing in Zeitschrift des Vereins deutcher Ingenieure. He stated "The more elastic the teeth are, the greater the error that can be permitted."

By the early 1920's high speed gears were experiencing "hammering" and flattening of the teeth curves. Their solution was to cut the teeth so accurately that no backlash or side clearance existed. They learned that due to the low elastic limit of the cast irons and bronzes the most reliable gear material was steel. Rolling mill pinions were tried with 0.3 percent carbon and were destroyed in a few months. When the steel was changed to 0.6 percent carbon they lasted several years. Lewis in a discussion on high speed gearing objected to the concept current at the time that speed and pressure were constant. He believed that it had been learned how tooth pressure should vary with speed. From experiments and discussion based on Lasche's paper it was believed that neither cycloidal nor involute curves provided the best results and an entirely new tooth form was proposed. By dividing the length of two working teeth into an equal number of parts, the amount of sliding could be determined.

Epicyclic: In Paris, France, the first successful use of planet gearing occurred in 1828. Alex B.W. Kennedy described epicyclic gear trains "… in them one or more wheels revolve about the fixed one, in such a way that points in these wheels describe different cycloidal curves…" In 1889, the American Machinist published on February 14th an informative article by Professor A.T. Woods on epicyclic gearing. Oscar J. Beale was to write in the same journal, July 1908, "This article is well nigh perfect."

The use of internal gearing for driving rolling mills was patented in England by W.N. Nicholson in 1858, to improve on the imperfect spur gears that were being used. (p.245)

In 1880 the American Society of Mechanical Engineers was conceived by a meeting of approximately eighty engineers in New York City on February 16th. In 1898 the National Bureau of Standards was founded in the U.S., and in Britain the National Physical Laboratory was established.

Bicycles: The history of self-propelled two wheel transport is pre 19th century. The first steerable two wheel bicycle the Draisine was introduced by the German, Freiherr Drais, in 1817. The year 1816 was known as the "Year without a Summer" due to the Mount Tambora eruption, as a result of the devastation a new mode of transport was needed. Various improvements took place over the next ninety years that supplied the technology for all future transport. This predecessor of the automobile and aircraft created a revolution in public transport. Innumerable designs of velocipedes, bicycles, tricycles, and multicycles had become available in the latter part of the 19th century. In the period 1890-1918 more than three thousand brands of bicycles were built. The invention of Starley's wheel in 1876 led to an amazing number of designs, as Sharp wrote "in utter ignorance of mechanical science". John Staley introduced the "Safety Bicycle" in England in 1875. The demand for personal transport all started with the bicycle. By 1896 the press was extolling the bicycle as "the greatest invention of the nineteenth century." The bicycle created an interest and unprecedented demand for personal transport and in doing so advanced the engineering and production of materials, bearings, and gears. Susan B. Anthony credited the bicycle with having done more to liberate women than anything else. The Census Bureau in 1900 noted "Few articles created by man have created so great a revolution in social conditions."

A Gear Chronology

The cycles required large quantities of mass produced gears, especially bevels, which led to rapid advances in both gear manufacturing and gear technology. Stresses and dynamics would now be better understood as would quantity manufacturing and quality control.

The growth in the bicycle industry was phenomenal. In the seven years following 1890 the number of manufacturers grew from 25 to 312. In 1895 1.2 million units were produced which gave new meaning to the term mass production. Pope Bicycles employed 500 in '88 and 3,800 in '99. At the Pope Company, Hartford, Connecticut, numerous small bevel gears were produced on a machine designed by H.C. Warren. The machine had two rotary cutters mounted on axes at an angle with each other. The blank rotated on its axis. The Fearnhead gear used a bevel gear attached the crank and hub axle. If these bevels could be accurately and cheaply machine cut it was possible that gears of this description would replace chains. Several compound drives, such as Hart's Gear and Devoll's Gear had been previously designed to avoid the use of a chain.

These inventions would create a previously unprecedented demand for high volume, improved quality and new designs of gears. In 1823 patents on differential gearing were issued to Asa Arnold of Rhode Island, Henry Houldsworth and Green in England. In 1829 the differential was "*invented* in France by Pasquier, patented in England by John Hannon, and by Starley in 1877. We also know the differential was used in China B.C., and has been re-invented several times. In England, James Starley invented the *"Ariel"* geared bicycle and in 1877 the differential for his *"Coventry"* tricycle as shown in figure 5-84.

Figure 5-84 The Cyclo gear used in *Starley's* differential tricycle

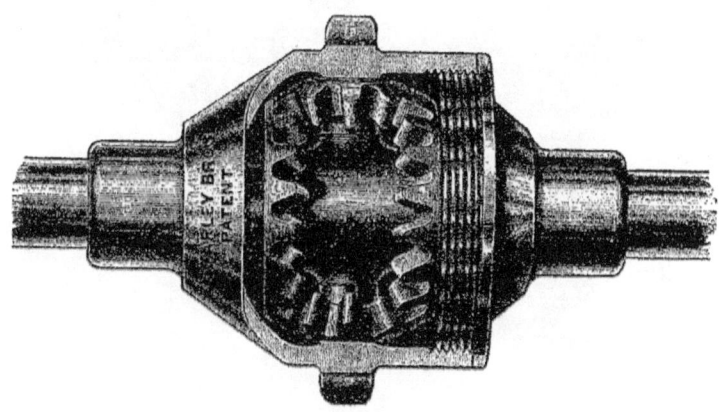

William Healy patented in 1895 a multi-speed bicycle hub gear in England as shown in figure 5-85. The *Healy Gear* used an epicyclic bevel gear with a ratio of 2:1, and was more compact than the Boudard gear shown in figure 5-88.

Figure 5-85 Healy Gear

Figure 5-86 Perry's Front Driving Gear similar to a lathe's back-gear.

Other gear designs were the Crypto figure 5-87, Boudard figure 5-88, the Platnauer figure 5-89, and Cycle Gear 5-90.

Figure 5-87 *Crypto* Front-Driver

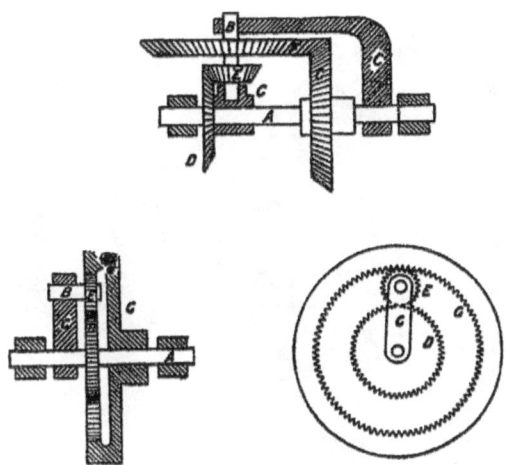

The Boudard Gear was the first of a number of compound driving gears that were used in combination with a large chain wheel in the rear hub. (Figure 5-86)

Figure 5-88 Boudard Gear

The Platnauer geared hub with fixed pinion figure 5-89 was not successful.

Figure 5-89 Platnauer Gear

Figure 5-90 The Cycle Gear Company's two-speed gear with an epicyclic train similar to the Cyclo.

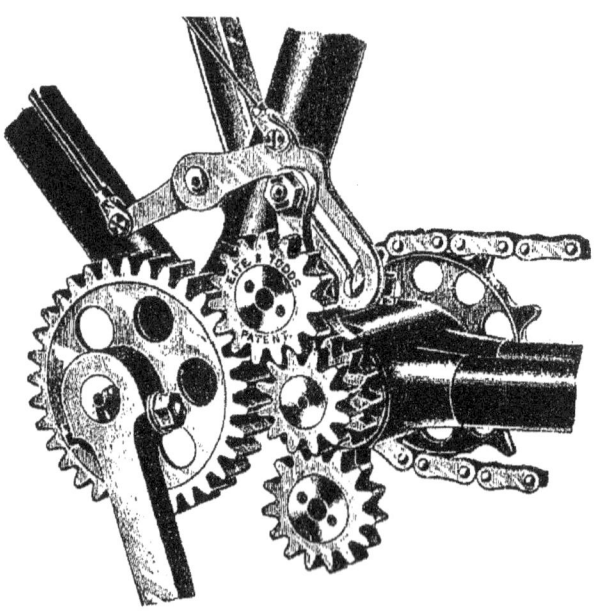

In 1896 "Bicycles and Tricycles", an important book on the industry by Archibald Sharp, was published in London and New York. (1977 reprint MIT Press). The book includes a chapter on gear engineering and involute, cycloidal and circular tooth forms. Sharp describes an important stage in the accelerating growth of mechanical engineering that was initially due to James Starley's invention of the tangent-spoked tension wheel in 1876 and his previous cycle designs. The disadvantage that applied to all bevels of the period was that they could not be made accurately and at low cost.

Automobiles: Between 1897 and 1914 the multi-cycle industry collapsed, to be superceded by the automobile. Many of the early automobile companies had been bicycle manufacturers, Opel in Germany; Humber, Riley, Rover and Morris in England, Winton, Willys, Pope, Peerless and Rambler in the U.S. Some automobiles contained features of the modern vehicle and used knowledge gained form bicycle designs. Lanchester's vehicle of 1895-6 had bicycle wheels and tires. The car's single cylinder fan-cooled five horsepower engine had an epicyclic gear box to provide direct drive in high gear, or low and reverse speeds. The circulation of oil lubrication was mechanized. The horizontally opposed pistons were geared together in such a manner so the crankshafts with separate flywheels ran in opposite directions.

The automobiles resulting from the new power sources would also be a major influence on gear development. The automobile industry can be said to have begun and stalled in Britain. In 1865 the "Locomotives on Highway Act", misnamed the "Red Flag Act", was introduced requiring self propelled vehicles to be preceded by a man on foot. The law was in effect until the so-called1896 "Emancipation Act", to the detriment of British industry. This act instituted a speed limit of 12 mph. which was raised to 30 mph. in 1904 and this speed limit would remain in effect until 1930. In the following year 1897 Frederick and George Lanchester built the first all-British car. The drive was through an epicyclic gearbox to a differential gear rear axle. They both were interested in evaluating and developing gears.

Based on Davenport's experiments with electric motors, Moses G. Farmer of Massachusetts bolted an electric motor onto a carriage in 1847, the first unofficial electric car. In 1891 the first official U.S. electric automobile was designed by William Morrison in Des Moines, Iowa, requiring less complex gearing.

In the U.S. in 1900 4,192 automobiles would be built, 1,681 were powered by a steam engine, 1,575 by an electric motor, and 936 with an internal combustion engine. At the turn of the century even airplanes were being designed with steam engines, and in the 1897 Avion 111 the engines were half the weight per horsepower of the internal combustion engine used by the Wright brothers. An automobile change speed gearbox was invented by H. Panhard in France in 1889. Another major advancement in automobile design was made in 1898 when Henry Timken and B. Heinelman patented a tapered vehicle roller bearing.

Flight. The Wright brothers were both bought safety bicycles and in 1896 introduced their own machines. They designed a self lubricating hub. When they achieved controlled and powered flight nine years later observers said it looked more like a bicycle than a bird. They used a bicycle to determine lift and drag and the forces on different wing shapes. They used a four cyclinder gasolene engine with an aluminum crankcase.

Glen Curtis also started with bicycles before becoming the main rival to the Wrights. He was fifteen years old and working for Eastman Kodak when he bought his first bike for $125. He competed and raced his bicycle. He then moved on to engines and motorcycles. He installed a V-8 engine in 1907 and was known as the fastest man on earth when he travelled at 136 mph on a one-mile track. His light weight and powerful engines put his talent and experience into designing aircraft. He was awarded a *Scientific American* prize for the first observed flight of one kilometer. The Wrights had flown longer distances but they were unrecognized because of a lack of witnesses. In France he set an airspeed record by using his bicycle racing experience. Diving to recover speed after the turns. In the U.S the bicycle would go into a decline, used by children until they could obtain a car. Since the 1970's however there has been increasing demand.

Lubrication: In the early part of the century lubrication was far from an art. Nathaniel Partridge obtained the British patent #3573 in 1812 for "A composition to prevent friction..." The heavy grease comprised a dry graphite lubricant mixed with animal fat. He obtained another patent #6945 in 1835 for an anti-attrition paste made mainly from olive oil and a solution of lime in water. For heavier duties such as lubricating "...cogs of teeth of wheels..." whale oil thickened with palm oil or tallow and with an additive of carbon from plumbago, black lead or soot was preferred. In 1849 patent # 12571 was issued in Britain to William Little for the use of a distilled mineral oil to lubricate machinery. The first mineral oils used for gear lubrication were sticky and resisted being displaced by the tooth pressures. Following 1860 petroleum based lubricants were to be more generally available. The transition from vegetable and animal oils was of very short duration and had an immediate impact. However, even in 1880 olive oil would be one of the most popular lubricants.

Oil and grease cups came into use in 1820, and in 1870 pressure feeding grease cups and box oilers. In the U.S. Albert Charles Pain patented in 1890 a force lube system, the oil traveled through small canals to the journals and bearings. Also in 1890 the Belliss and Morcom Company introduced force feed lubrication in England. With the introduction of high speed steam engines lubrication became critically important.

Professor Thurston, whom we have previously mentioned his lubrication and friction lectures and important 1885 book (p.145). He was an early tester of lubricants building a pendulum testing machine and provided coefficients of friction under field operating conditions for a wide range of lubricants.

The general lack of knowledge on lubrication is evidenced by the letter to the editor of "Power" in January, 1885, requesting details on the reported saving of fifteen to twenty tons of coal a month using improved lubrication.. "We have for some years been trying to persuade the average lunkhead engine-owner and pig-headed "engineer'...there is no economy in first sloshing an engine all over, inside and out, with oil and then let it run dry and cutting..."

The mechanism of fluid film lubrication was developed from experimental work and mathematical analysis, Professor Osborne Reynolds reported in 1885 to the British Association in Montreal his theory of fluid-film lubrication. In 1888 he was the first professor of engineering in Manchester, England, and a Royal Society gold medalist. Reynolds studied the dynamic state of fluids and lubrication. The impact of hydrodynamic lubrication would not be fully utilized for another twenty or so years. The Reynolds Number originated from his name. Proceedings of the British Institute Mechanical Engineers 1883, 1885 and 1888 by Professor Tower, and researches by Professor Reynolds and later by Oscar Lasche in 1903 were aimed at solving the problem of bearing lubrication at the higher speeds required by machine tool spindles.

The role of viscosity in fluid –film lubrication was studied by W. Bridges Adams in Englan0d and Gustav A. Hirn in France with different conclusions. Reynold's understood the need to measure viscosity and its relationship to temperature. He stated in a student paper to the Institution of Civil Engineers in 1886, "...wherever hard surfaces under pressure slide over each other without abrasion, they are separated by a film...". Once Reynold's equations had been solved for bearings in the first quarter of the 20th century they would be applied to the lubrication of gears.

John Goodman, later Professor of Mechanical and Civil Engineering at Leeds University, identified the existence of a lubricating film, and for the first time provided a quantitative measurement of its magnitude. His micrometer measurement was able to measure to 0.0001 inch.

New Gear Applications: Edwin Maw, Liverpool, England supplied a three roll sugar mill to R.W. Meyer in Maui in 1843 using bevel gears shown in figure 5-89. A bevel gear at the lower end of a vertical drive shaft drove the bevel gear of the top roll. The shaft was supported in a brass-bearing socket. The illustration shows that the gears now available could no longer be considered crude. The gears were durable and relatively precise. They were a major advance over the previous century.

Figure 5-91 1843 Bevel Gears

In California in 1868 Philander Standish built the twelve horsepower Mayflower, a steam driven rotary plow. The plow had a vertical boiler, horizontal cylinder, and a rod-and-crank to the first of several stages of gears. The final stage drove a ring gear at the rear wheels. The plow was capable of plowing five acres an hour. The growth in power driven agricultural machinery was encouraged by President Lincoln. In his address to the Wisconsin Agricultural Society in 1859 Lincoln stated: "… must … plow better than can be done with animal power. It must do all the work as well, and cheaper, or more rapidly."

Typical of these self-propelled steam engines was the tractor introduced by the J.I. Case Company in 1870. From the PTO to the rear wheels was a train of seven gears which further increased the demand for reliable mass production gears.

Figure 5-92 J. I. Case Self-propelled Steam Tractor Circa 1880

In the latter part of the 1800's, demand also increased for self-propelled threshing machines, three thousand were built in 1890. The typical transmission contained cast iron spur and bevel gears. A friction clutch connected the crankshaft to a pinion which engaged an intermediate gear. This gear meshed with a large compensating gear on a countershaft. On each end of this countershaft was a pinion driving a main gear at each drive wheel.

Another important gear market arose with the demand for newspapers and the need for rapid printing. The 1840 census indicated a population increase of thirty-two percent and a newspaper circulation increase of 187 percent. The answer to increased production was found in the double-cylinder rotary press. For the first time both sides could be printed and the ink evenly and quickly spread. Between 1828 and 1847 Robert Hoe and his son Richard developed the cylinder press that eliminated the flat bed press. Steam driven, with ten cylinders and weighing forty tons, they could turn out 20,000 impressions an hour. Each cylinder required an eight spoke gear in excess of six foot meshing with the next cylinder. Large gear trains with as many as four gears were also required on each cylinder.

Herrmann and Weisbach considered various gear mechanisms such as the innovative reverse gear shown in figure 5-93.

Figure 5-93 Hermann and Weisbach's Reverse Gear Circa 1890

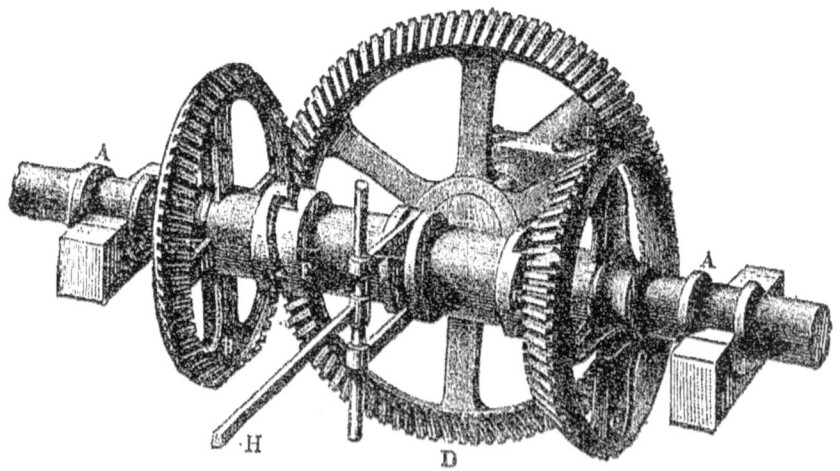

These developments and particularly the advances in available power sources meant that gear technology and production would be accelerated into the 20th century.

Index of Individuals

A

Acheson, Edward G., 232
Adams, W. Bridges, 329
Adamson, David, 401
Adelard of Bath, 51
Aeppli, Albert, 484
Agricola, Georgius (George Bauer), 106
Ahmose (Egyptian mathematician), 32
Aida, T., 367
Aiken, Herrick, 178
Aiken, Howard, 515
Airey, J., 376
Airy, George Biddell, 298
Albert, W. A. J., 196
Alden, George I., 463
Alford, L. P., 375
Allen, C. M., 476
Allen, S. K., 390
Almen, J. O., 509
Amontons, Guillaume, 83
Ampère, André Marie, 181
Anderson, H. N., 257, 444
Anderson, Robert, 183
Andrew, H. J., 368
Anthony, Susan B., 321
Apollonius of Perga, 35
Archbutt, L., 500
Archimedes (Greek mathematician), 34, 67
Aristotle (Greek philosopher), 41
Arkwright, Sir Richard, 156
Armstrong, Edwin H., 335
Arnold, Asa, 321
Aryabhata, 48
Attia, A. Y., 401
Augustus (Roman emperor), 54
Ayton, William Edward, 203

B

Babbage, Charles, 160, 204, 514
Babbit, Isaac, 190
Babinet, Jacques, 203
Bach, C. E. Prof., 470
Bacon, 51
Baekeland, Leo Hendrik, 370
Baldewin, Eberhart, 96
Baldridge, Malcolm, 344
Balzer, 381
Barbaro, Danile, 55
Barclay, Henry, 229
Bardeen, John, 518
Bardons, 253
Barlow, 409
Barsanti, 185
Barth, Carl G., 318
Bartlett, J. M., 392, 439
Bartlett, W. H. C., 161
Basov, Nikolai, 524
Bates, W. C., 386, 400, 409
Baud, R. V., 402
Baxter, 497
Beale, Oscar J., 272, 320
Beauchamp Tower, 197, 488

Bede, 49
Beilby, Sir George, 404
Belanger, 306
Belfield, Henry, 263
Bèlidor, Bernard Forest de, 121, 129
Bell, Alexander Graham, 480
Bell, Daniel, 338
Bement, 218, 244, 313
Benedict, 495
Bentham, Samuel, 149
Benz, Karl Friedrich, 185
Bergmann, Torbern Olaf, 127
Bernardos, N. V., 184
Bernoulli, Jacques, 83
Bernoulli, Johannes, 83
Berthoud, Ferdinand, 144
Bessemer, Sir Henry, 188–89
Besson, Jacques, 99
Bétancourt, 159
Bháscara, 51
Bierbaum, Christopher H., 368
Bilgram, Hugo, 256, 263, 479
Bion, Nicholas, 101
Bird, W. W., 375
Biringuccio, Vannoccio, 73
Biruni, al-, 35, 51
Bishop, 158
Black, Harold S., 335
Black, Paul H., 403
Blackhaus, Louis, 164
Blanc, H., 149
Blanchard, Thomas, 211
Blenkinsop, John, 177
Blok, Harmen, 496
Bodmer, Johann Georg, 207
Bond, George M., 201
Boner, C. J., 498
Borgis, Tycho, 160
Borsoff, V. N., 496
Bostock, F. J., 336
Bottcher, 392, 438
Boulton, Matthew, 133, 172
Box, Thomas, 300, 316
Boyle, John, 92
Bragg, William Henry, 352
Brahe, Tycho, 79

Brainard, Horace E., 520
Bramah, James, 214
Bramah, Joseph, 226
Bramley, 395
Branca, Giovanni, 115
Brandenberger, 442
Brattain, Walter Houser, 518
Brayshaw, Shipley N., 352
Brayton, George, 186
Breckenridge, Prof. l. P., 195
Brewster, Sir David, 287
Bridges, Jonathan, 233
Briggs, Henry T. R., 80
Brindenburger, 394
Brindley, James, 152
Brinell, Dr. Johann August, 363
Brown, David, 336, 368, 409, 447, 465
Brown, Henry, 306
Brown, James S., 234
Brown, J. R., 201, 224
Brown, Sylvanus, 150, 234
Brugger, H., 352
Brunel, Sir Marc Isambard, 227
Brunelleschi, Filippo, 70, 102
Buchanan, Joseph, 178
Buchanan, Robert, 244, 301
Buckingham, Earle, 303, 401, 409
Buckle, William, 212
Bullard, E. P., 263
Bundy, Francis, 378
Buot, Jacques, 100
Burgess, F., 473
Burgi, Jost (Justiust Byrgius), 80
Bush, Vannevar, 515

C

Caesar, Julius, 40
Cajetano, Aureliano, 140
Cajetano, David, 140
Cajori, Florian, 33
Cal Lun, 53
Camus, Charles Etienne Louis, 286
Caous, Solomon de, 115
Capek, Karel, 525
Cardan, Jerome (Girolamo Cardano), 88
Cardwell, D. S. L., 93

Carlson, Chester, 516
Carlyle, Thomas, 129
Carnegie, 190
Caruthers, Wallace Hume, 372
Carvill, G. C. and H, 289
Castellani, G., 403
Cauchy, Baron Augustin Louis, 160
Cellini, Benvenuto Prof., 75
Cellini, Prof. B., 38
Celsius, Anders, 131
Chace, N. B., 452
Chambon, M., 432
Chao, 497
Charles V, 100
Chasles, Micel, 306
Chateliere, Henry le, 345
Chaulnes, Duc de, 154
Cheng, 497
Christin, Jean Pierre, 131
Chuchill, Sir Winston, 516
Church, 389
Clark, Victor S., 139
Clavius, Christophorus, 40
Clement, Joseph, 144–45, 204, 209–10, 215–16, 225, 287–89
Clerm, 276
Cohen, M. R., 55
Coker, Prof. E. G., 400
Collier, 293
Colvin, 358, 396, 419
Combe, 168
Comley, John, 257
Cone, Samuel I., 389
Conradson, Conrad, 263
Cook, 182
Cooper, H. S., 499
Cooper, John, 318
Cooper, Peter, 179, 233
Copernicus, 40
Corliss, George H., 251
Cort, Henry, 124–25
Cotton, 259
Coulomb, Charles Augustin de, 128
Cox, John L., 345
Coy, 406
Cromwell, J. Howard, 397

Croning, Johannes, 352
Crook, A. W., 495
Ctesibius (Greek engineer), 41, 55
Cubitt, William, 165
Cugnot, Nicholas Joseph, 135
Cunningham, Dr. Fred, 520
Cutter, Prof. Lawrence C., 399

D

Daedalus, 38
Daimler, Gottlieb, 185–86
Damuell, V. R., 499
Darby, Abraham, 124
Darling, Samuel, 233
Dauthiau, 153
Davenport, Thomas, 182, 326
Davidson, Ellis A., 292
Davis, Ernest F., 354
Davis, G. H. B., 488
Davis, William, 208
Day, Percy. C., 385
Dean, E. W., 488
Deeley, R. M., 500
Deleeuw, A. L., 374
Dengg, C., 260
Denny, 475
Deplangue, 229
Derihon, 363
Desargues, Girard, 89, 105
Descartes, Renè, 81
Devol, George C., 525
Devoll, 321
Diderot, Denis, 100, 123
Diefendorf, W. H., 370
Diesel, Christian Karl Rudolph, 186
Din, Taqi al-, 111
Diophantus (Greek mathematician), 49
Doaln, Thomas J., 402
Dominy, Nathaniel, 164
Dondi, Giovanni di,' 62, 94
Donkin, Bryan, 230
Dowson, D., 495, 497
Drabkin, E. I., 55
Drais, Freiherr, 320
Drummond, R. S., 456
Dudley, Darle, 163

Dudley, Dud, 73
Dufraine, 496
Duke of Chou, 59
Dürer, Albrecht, 68, 77–78, 130, 284, 294
Duryea, Charles E., 335
Duryea, Frank, 335
Duval, Gideon, 145

E

Eberhardt, Henry J., 432
Eckert, John Presper, 517
Edison, 71, 334
Edward 1, 47
Edwards, C. A., 368
Elias, 182
Elizabeth the First, 47, 80
Ellyott, 73
Emerson, 127
Emery, A. H., 376
Engler, 505
Eratosthenes, 34
Ernault, H., 438
Ertel, A. M., 494
Euclid, 34, 90
Euler, Leonhard, 88, 91–92, 121, 129, 137, 302
Evans, Oliver, 116, 134, 169–70, 240
Exiguus, Dionysius, 49
Extot, Manouri d,' 168
Ezoe, S., 456

F

Fahrenheit, Daniel Gabriel, 131
Fairbairn, William, 187
Fairfield, H. P., 375
Faraday, Michael, 180–81, 191–92
Farish, Reverend William, 285
Farmer, Moses G., 326
Fässler, Albert, 467
Fast, Gustave, 404
Faulconer, 236, 238
Fearnhead, 321
Fellows, E. R., 268, 383, 428, 456
Ferro, Scipione da, 88
Fibonacci, Leornado, 52
Field, Joshua, 151
Firth, M., 233

Fischer, Friedrich, 234
Fischer, Johann Conrad, 192
Fitch, James, 173
Fitch, Stephen, 212
Flanders, Ralph E., 398
Floe, Carl, 360
Flowers, 501
Foeppl, 200
Foley, Francis B., 345
Fontaine, Hippolyte, 183
Foote, 253
Ford, Henry, 186, 207, 507
Forq, Nicholas, 144
Forrester, Jay W., 519
Fouche, Edmund, 185
Fourier, Joseph, 181
Fourneyron, 168
Fox, James, 144, 215, 242
Franken, Peter Al, 480
Franklin, Benjamin, 173
Frederick the Great, 129
Froissart, 95
Fry, Dr. A., 360
Fulton, Robert, 172

G

Gaius Plinius Secundus the Elder, 42
Galileo, 76, 82, 88, 121, 130, 139
Ganschow, William, 406, 408
Garay, Blasco de, 172
Gardner, Frederick M., 238
Gascoigne, William, 117
Gaudard, Lucien, 184
Gay, Ira, 234
Geier, Frederick A., 208
Georgi, C. W., 505
Georgio, Francesco di, 55
Geyellin, 319
Gharibal-Hadith, Kitab, 53
Gibbs, J. D., 184
Gille, Bertrand, 68
Glavet, 251
Gleason, A. C., 348–49
Gleason, James E., 396, 423, 439
Gleason, William, 253
Goldschmidt, Johann, 185

Goodman, John, 329
Goodrum, Thomas, 235
Gordon, L., 167
Gould, Ezra, 244
Gould, R. Gordon, 524
Gramme, Zénobe Thèophile, 183
Grant, George B., 264
Grassi, Fabrizzio, 521
Gray, George A., 208
Green, 321
Greening, 409
Gregory, R. W., 404
Gregory XIII (pope), 40
Gribben, Walter, 309, 390
Griffith, A. A., 351
Grob, Benjamin, 445
Grosse-teste, 51
Grubin, A. N., 495
Guericke, Otto von, 116
Guest, James J., 464
Guillaume, 83
Gumbel, Ludwig K. F., 491
Gunter, Edmund, 80
Gutierrez, Manuel, 155

H

Hâchette, Jean Nicholas Pierre, 159
Hadfield, Sir Robert Abbot, 189
Hagen-Thorn, 259
Haitham, Alhasan ibn al-, 51
Hall, John, 222
Hall, Joseph, 125
Halsey, F. A., 368
Hannon, John, 321
Hardy, William Bate, 494
Harris, Stephen, 404, 487
Hart, Gilbert, 230–31
Hartness, James, 268
Hasin, Habash al-, 50
Hawkins, John, 88, 285–86, 288
Healy, William, 323
Heath, J. M., 192
Heath, T. L., 35
Hedley, William, 178
Heinecker, 264
Heinelman, B., 327

Henry, Joseph, 180
Henry IV, 79
Henry VII, 47
Hephaestus, 37
Herbert, Edward G., 366, 376
Heron of Alexandria, 41, 55–57, 174
Herrman, Prof. Gustav, 256
Herschel, Sir John Frederick William, 203
Herschel, W. H., 505
Hersey, Mayo D., 488
Hertz, Heinrich Rudolph, 313
Hiersig, Professor, 495
Higginson, G. R., 495
Hill, Donald, 64, 93
Hindley, Henry, 152, 310
Hipparchus, 35
Hirn, Gustav A., 329
Hiscox, Gardner D., 468, 509
Hiyya, Abraham bar, 51
Hoe, Richard, 332
Hoe, Robert, 332
Hoecke, G. V., 78
Hoersch, V., 405
Hoffman, J. G., 267
Hollerith, Herman, 204, 514
Holtz, William, 225
Holtzapffel, 160–61, 163, 187–88, 228–29, 249, 314, 444
Holtzapffel, John Jacob, 160
Holtzer, Jacob, 193
Holz, Fred, 208
Holzschuler, 69
Homer (Greek poet), 37, 39
Honnecourt, Villard de, 64
Hooke, Dr. Robert, 92, 118
Hoover, Herbert, 106
Hopps, 504
Hornblower, Josiah, 132
Houghton, E. F., 190
Hou Han Shu, 48
Houldsworth, Henry, 321
Howe, Henry, 345
Howe, William Frederick, 174
Huaiwen, Qiwu, 47
Humpage, Thomas, 462
Hunt, C. W., 383

Huntsman, Benjamin, 127
Hutton, F.R., 214
Huygens, Christian, 88–89, 116–17, 119
Hyde, J. H., 501

I

Ibn al-Awwâm, 46
Ibn Sida, 46
I-Hsing (Buddhist monk), 52
Imison, John, 140
Isfahânî, Muhammad Abî Bakr ar-Râshidî al-ibarî al-, 51
Ishibashi, A., 456
Istakhri, al-, 65

J

Jackson, 389
Jacobi, 182
Jacobs, Charles B., 231
Jacquard, Joseph-Marie, 514
Jameson, Dr. Alexander, 160
Jarchow, 495
Jazari, al-, 63–64
Jefferson, Thomas, 119
Jenks, Joseph, 70
Jerome, Chauncey, 242
Johansson, C. E., 202, 480
Johnson, Robert, 221
Jominy, Walter E., 367
Jones, Franklin, 415
Jones, William, 76, 128
Joule, James Prescott, 181
Jublot, R. M., 499

K

Kaestner, Abraham Gotthelf, 92
Karsten, 187
Kashi, Ghiyath al-, 76
Keller, Joseph, 532
Kelly, 495
Kelly, William, 189
Kempelen, Baron, 174
Kennedy, Alex B. W., 198, 309, 319
Kenward, Cyril Walter, 526
Kepler, Johannes, 514

Ketcham, W. J., 360
Khwarizimi, al-, 50
Kilby, Jack, 517
King John, 47
Kingsbury, Albert, 405
Kirkaldy, David, 194, 303, 314
Klein, Prof. J. F., 250, 256
Knudsen, W. S., 335
Koehler, W., 499
Köller, Franz, 192
Köller, Josef Jacob, 192
Kraus, G., 352
Krupp, Friedrich, 377
Kunzel, 191
Kyeser, Konrad, 68

L

Lagrange, Joseph-Louis, 121, 129–30
la Hire, Phillippe de, 84
Lanchester, Frederick, 264
Lanchester, George, 326
Landop, Christopher, 152
Langen, 185
Lanz, 159
Laplace, 121
Lapointe, John N., 444
Lasche, Oscar, 315, 319, 329, 488
Lavroff, 191
Lawson, 259
Leblond, Richard K., 208
Legendre, Adrien Marie, 129
Leibniz, Gottfried Wilhelm, 82–83, 118, 121, 514
Leland, Henry M., 235
Le Lievre, 145
Lenoir, Jean Joseph Etienne, 185
Leonardo da Vinci, 85–86
Lepel, 360
Le Rond, Jean, 123
Leupold, Jacob, 105, 121–22
Leutwiler, O. A., 371
Levassor, M., 186
Levi, 191
Lewis, F., 244
Lewis, Wilfred, 196, 198, 340, 381, 401, 471
L'Hopital, Guillaume De, 83

Liebergold, A., 379
Liebig, Baron Justus F. von, 187
Lightfoot, Peter, 96
Lilius, Aloysius, 40
Lillie, 182
Lincoln, President, 330
Lister, C. A., 469
Litchfield, Norman, 384
Little, William, 328
Litvin, Faydor L. Professor, 468
Liu Hui, 49
Lodge, William, 208
Logue, Charles H., 313, 340, 397
Louis XIV, 100
Loveless, 409
Lucas, Samuel, 187
Ludblad, Rink, 378
Lundberg, 405–6

M

Mac Cord, 152, 250, 262, 279, 281, 283, 291, 296, 307–8, 392
Macquer, Pierre Joseph, 128
Maiman, Theodore, 524
Manley, L. W., 499
Mansfield, A. K., 300
Marcus, Siegfried, 185
Martel, 200
Martens, Adolph, 190
Martin, H. M., 490
Martini, Francesco di Giorgio, 103
Marx, Guido H. Professor, 399
Matlock, 488
Mauchley, John, 517
Maudslay, Henry, 150
Maw, Edwin, 330
Maximilian the First, 78
Maxwell, James Clerk, 273
Mayall, T. J., 229
McDonough, James C., 519
McMullan, O. W., 353
McQuaid, H. W., 353
Meikle, Andrew, 138
Merrick, S. V., 216
Merritt, H. E., 475
Mesta, George, 268

Meyer, R. W., 330
Meysey, 73
Michell, Anthony George Maldon, 405
Miller, F. L., 494
Miller, John, 372
Minorsky, Nikolai, 273
Mohs, Friedrich, 199
Moinet, 95
Moissan, Henri, 232
Mönch, Philipp, 69
Monge, Gaspard, 130, 159
Monneret, M., 392, 394, 438, 442
Moore, Gordon, 521
Morin, Arthur Jules, 197
Morris, Dr. D. K., 469
Morrison, William, 326
Morse, Stephen A., 276
Moseley, 198
Mougey, H. C., 493
Moureau, 496
Mouton, Abbé, 119
Moxon, John, 119
Mueller, George, 208
Mumford, Lewis, 45
Munro, R. G., 404
Murâdi, al-, 61
Murdock, William, 134, 146, 149, 220
Murray, Matthew, 144, 149, 215, 220
Mûsà, al-Hasan Banû (three brothers), 52
Mushet, David, 127, 192
Mushet, Robert Forester, 192
Musschenbroek, Petrus van, 139
Myers, 406

N

Napier, John, 79, 117
Napoleon III, 207
Nasmyth, James, 151, 216
Natrus, 137
Neergaard, Leif Eric de, 516
Neilson, J. Beaumont, 187
Nemorarius, 51
Neumann, John von, 517
Newcomen, Thomas, 131
Newton, Isaac, 82, 96, 119
Newton, Sereno, 289, 302

Nicholas, Emperor, 182
Nicholas of Cusa, 87
Nicholson, W. N., 299, 320
Niemann, Prof. G., 389
Norris, H. M., 374
North, Co. Simeon, 221
Norton, Charles H., 236, 459
Norton, Franklin B., 231, 460
Norton, W. P., 214, 412
Novikov, M. L., 395
Noyce, Robert, 517
Nuttall, R. D., 341, 383, 386, 439, 510
Nyquist, Henry, 515

O

Oberg, Erick, 350–51, 362–63, 415
Oberg, Erik, 350, 415
Octrue, Michel, 369
Oda, S., 367
Olivier, Th., 304
Olsen, Tinius, 196–97, 488, 504
Orpheus, 38
Osmond, Floris, 189, 195
Otto, Nikolaus A., 185
Oughtred, Dudley William, 73
Oxford, C. J., 376

P

Pacioli, Luca, 76
Page, Dr. Charles, 182
Pain, Albert Charles, 328
Palmer, J. R., 201
Palmgren, Arvid, 405
Panhard, H., 326
Panhard, René, 186
Papin, Dr. Denis, 116
Pappus of Alexandria, 49
Parent, A., 84
Parker, John, 418
Parkes, Alexander, 191
Parkhurst, E. G., 212
Parkinson, Joseph, 226
Parks, Edward H., 274
Parr, P. H., 500
Parsons, C. A., 174, 432
Parsons, John Thoren, 519

Partridge, Nathaniel, 327
Pascal, Blaise, 82, 89–90, 118, 514
Pasquier, 321
Patterson, 218
Payne, John, 112
Peale, Franklin, 218
Pearson, 285
Pease, William M., 519
Pellos, Francisco Bel, 76
Pentz, A. D., 309, 390
Perry, John, 203
Pettus, John, 47, 119
Pfauter, Hermann, 264
Phillip VI, 68
Philo of Byzantium, 41
Pickard, James, 145
Pixii, Hippolyte, 182
Platt, Hugh, 74
Pliny. *See* Gaius Plinius Secundus the Elder
Pliny the Elder, 47
Plumier, Charles, 140
Poinsot, 304, 306
Poisson, Siméon Denis, 196
Polhelm, Christopher (Polhammer), 101
Poliakoff, R., 375
Pollio, Marcus Vitruvius, 54
Polly, 137
Poncelet, Jean Victor, 207, 289
Poole, J. M., 234
Pope, Colonel Albert, 335
Porta, Giambattista della, 76
Potemkin, G. A., 149
Power, Henry, 118
Pratt, Francis, 224
Predki, Professor, 497
Preis, 394, 443
Presper, John, 517
Priestley, Joseph, 127
Prokhorov, Alexander, 524
Ptolemaeus, Claudius (Ptolemy), 35, 38
Pulson, Sven, 230

R

Ramelli, Agostini, 68, 109–11, 137
Ramsden, Jesse, 142, 154, 275
Ramsden, John, 142

Rankine, William J. M., 196, 303, 313–14, 401
Ransome, F., 229
Rashid, Harun al-, 53
Reason, 487
Réaumur, René-Antoine Ferchault de, 125
Reaves, R. H., 515
Recorde, Robert, 78
Redtenbacher, 168
Redwood, Boverton, 505
Regnault, M. V., 187
Rehé, Samuel, 154
Reid, Thomas, 287
Reinecker, Jüngst J. E., 264
Renk, Johann, 162
Rennie, George, 144, 197, 315
Rennie, John, 146
Reuleaux, Franz, 199, 297
Reynolds, Osborne, 329, 488
Richard of Wallingford, 94
Richards, Prof. C. B., 314
Richardson, Henry, 232
Riley, Professor, 473
Ripper, Prof. William, 375
Robert of Chester, 50
Roberts, Richard, 144, 151, 153, 160, 215, 217, 241
Roberts-Austen, Sir William Chandler, 189
Robinson (doctor), 248, 309
Rochas, Alphonse Beau de, 185
Rockwell, Stanley P., 365
Roe, Joseph W., 218
Roebuck, John, 124
Roemer, Olaus, 88
Röemer, Olaus, 89
Rogers, William A., 201
Rolt, F. H., 483
Roosevelt, Nicholas, 173
Root, Elisha, 224
Rose, Joshua A., 208
Rosenberg, Ralph H., 387
Roser, E., 470
Royce, Frederick Henry, 205
Rudolff, Christoff, 77
Rumford Count (B. Thompson), 130
Rumsey, James, 173
Ryder, 505–6

S

Saari, O. E., 394
Sandland, 366
Sang, Prof. Edward, 261
Sangallo, 105
Sasson, Bernard, 532
Sasson, Hans Ernst, 532
Saurer, Adolphe, 478
Sauveur, 190, 345, 350
Savage, Michael, 406
Savery, Thomas, 115, 131
Savile, Henry, 80
Saxton, Joseph, 257, 299
Schawlow, Arthur, 524
Schenker, Max, 515
Scheyer, Emmanuel, 514
Schicht, 394, 443
Schickard, Wilhelm, 514
Schickhardt, Heinrich, 105
Schiele, Christian, 266
Schieler, 46
Schlesinger, G., 437
Schlichthporlein, 456
Schlütter, 107
Schneck, R. B., 356
Schoop, 379
Sebohkt (monk), 50
Selden, George B., 186
Sellers, F., 375, 469
Sellers, William, 218, 313, 381–82
Senot, 141
Shang Gao, 33
Shannon, Claude Edward, 515
Sharp, Archibald, 312, 325
Sharpe, Lucien, 222
Shaw, John, 190, 532
Shih Huang-ti, 33
Shockley, William B., 518
Shore, 363–64
Shorter, A. E., 358
Showart, Dr. Walter, 480
Siemens, Sir Charles William, 184
Singer (doctor), 379
Sisco, A. G., 126
Slater, Samuel, 151

Smeaton, John, 122, 136, 146, 152
Smiles, Samuel, 147, 161, 206
Smith, Bill, 343
Smith, C. S., 118
Smith, D., 375
Smith, J. Kent, 346
Smith, L. L., 384
Sohne, 394, 443
Sokolov, 367
Solomon (king of Israel), 37
Sommerfeld, 490
Song Yingxing, 75
Sorby, Henry Clifton, 195
Southcombe, 494
Spencer, Christopher, 212
Sponaugle, Lloyd B., 516
Staley, John, 320
Standish, Philander, 330
Stanley, 358, 396, 419
Stansfield, 191
Stanton, Dr. T. E., 501
Starley, Dr. James, 322, 325
Stephens, Arison P., 444
Stephenson, George, 161
Steptoe, John, 208
Stevens, Col. John, 173
Stevin, Simon (Stevenus), 78
Stewart, Arthur L., 439
Stiffel, Michael, 78
Stimson, 182
St. John, Everet, 340
Stone, J. W., 233
Storm, John, 360
Strada, 68, 114
Stradanus, Johannes, 112
Stribeck, Richard, 315
Sturgeon, William, 181–82
Sunderland, Sam, 256, 427
Sung, James C., 378
Sung Ying-Hsing, 85
Susskind, Alfred K., 519
Su Sung, 60
Swasey, Ambrose, 260, 279

Swedenborg, Emanuel, 126
Swinburne, James, 370
Sykes, 430
Symington, William, 172

T

Taccola, Mariano di Jacobi detto, 67
Talleyrand, 119
Tanaka, S., 456
Tartaglia, Niccolò, 78
Taylor, Frederick Winslow, 193, 373
Telford, Thomas, 122
Tesla, Nikola, 184
Thales, 33
Theophilus, 47–48, 54, 119
Thiout, Antoine, 97, 140
Thomas, H., 405
Thomas, Sidney, 189
Thompson, 432
Thompson, Benjamin, 130
Thomson, Elihu Prof., 184
Thornton, Dr. William, 71
Thurston, Professor, 328, 379, 471
Timken, Henry, 327
Timoshenko, S., 402
Todhunter, 285
Torriano, Juanelo, 100
Tower, 197, 488
Townes, Charles H., 524
Townsend, 406
Tredgold, Thomas, 303
Trevithick, Richard, 134, 175
Ts'ai Lun, 38
Tschernoff, Dimitri, 199
Tubal-Cain, 36
Turner, Prof. Thomas, 363
Tuscany, Grand Duke of, 130

U

Uhlmann, Max, 393
Unwin, Professor, 312
Uqlidisi, Abu al-Hasan Ahmad ibn Ibrahim al-, 54

Ure, Andrew, 160
Utts, J. A., 439

V

Valturio, Roberto, 104
Van Zyl, J., 137
Vaucanson, 98, 141, 155
Vaucanson, Jacques de, 98, 141
Vernier, Pierre, 117
Viall, Richmond, 236, 459
Vick, Henri De, 95
Vigevano, Guido da, 68
Volta, Alessandro, 180

W

Waite, H. S., 340
Walker, Dr. Harry, 403
Walker, E. R., 196
Walker, H. C., 277
Walkley, 182
Wallis, John, 81
Wallwork, Henry, 472
Wandersee, John, 346
Washington, George, 138, 173
Watson, J. M., 353
Watt, James, 122, 132–33, 172
Wayland the Smith, 48
Weibull, W., 351, 405
Weisbach (professor), 199
Wellauer, E. J., 495
Wells, 494
Wentorf, Robert, 378
Westinghouse, George, 184, 476
Wheaton, James, 233
Whitaker, A. V., 497
White, James, 239, 285, 289
White, Maunsel, 193, 373
Whitelaw, James, 234
Whitney, Eli, 221
Whitworth, Joseph, 151, 200, 205, 265, 444
Wickenden, 161
Widman, Johannes, 78
Wiener, Norbert, 515

Wildhaber, Ernest, 395
Wilhelm IV, 96
Wilkinson, David, 151
Wilkinson, John, 147
Wilkley, J. E., 499
William III, 116
Williams, Harvey D., 384
Williams, John, 160
Williamson, D. T. N., 533
Williamson, Joseph, 153
Willis, John, 92
Willis, Robert, 91, 248, 285, 298
Willys, John, 335
Wilm, Dr. Alfred, 353
Wilson, Charles H., 365
Wilson, W. M., 476
Winthrop, John, 129
Wöhler, A. Z., 196
Wöhler, Friedrich, 194
Wolf, H. R., 493
Woods, A. T., 320
Wren, Christopher, 90
Wright brothers, 326–27, 335
Wüst-Kunz, Caspar, 450

Y

Young, Dr. Thomas, 195, 287, 289

Z

Zaretsky, 406
Zener, Clarence, 379
Zeno of Elea, 33
Zerk, Otto, 489
Zhang Heng, 59
Zimmerman, Johann, 208
Zonca, Vittorio, 111–12
Zorn, Dr. Herman, 498
Zu Chongzhi, 49

Index of Companies

A

A. Friedr Flender GmbH, 389
A. Harper Sons and Bean Ltd., 465
Alfred Herbert Ltd., 465
American Emery Wheel Works Co., 464
American Machine and Foundry Co., 526
American Steam Carriage Co., 176
Ames Manufacturing Co., 212
André Citroen & Co., 387
Arma Corp., 532
ASEA, 378
Automatic Machine Company, 412

B

Balzer, 381
Barber Coleman, 283
Bardons, 253
Bardons & Foote, 253
Barnes, 209, 520
Batelle, 367, 369
Bell Co., 533
Belliss and Morcom Co., 328
Bement & Dougherty, 218, 313
Benz Patent Motor Co., 185
Bessemer, 189
Bethlehem Steel Works, 193
Beyer Peacock Co. Ltd., 464
Bickford Drill & Tool Co., 374
Bilgram Co., 268
Blanchard Machine Co., 464

Boston Gear Works, 366
Boulton and Watt, 133, 212, 220
Boverton Redwood, 505
Bramah Works, 288
British Abrasive Wheel Co. Ltd., 464
Brown and Sharpe Mfg., 117, 155, 204, 222, 225, 233, 235–36, 243, 247, 340, 364, 388, 397, 418–19, 459
Brown Boveri, 360
Bryant Chucking Grinder Co., 464
Buick Motor Car Co., 356
Bullard, 253, 263

C

Cadillac and Lincoln Motor Co., 235
Cadillac Automobile Co., 236
Canton Iron Works, 179
Carbide and Carbon Chemical Co., 498
Carborundum Co., 232
Cardington Locomotive Works, 218
Case J. I., 331
Caterpillar Tractor Co., 456
C. E. Wüst, 450
Chas. H. Besly and Co., 464
Chelyabinsky Tractor Works, 445
Chevorlet Co., 335
Chrysler Corporation, 367
Churchill Gear Co., 527
Churchill Tool Co. Ltd., 465
Cincinnati Gear Co., 341
Cincinnati Grinder Co., 465

Cincinnati M/c (formerly Cincinnati Screw & Tap), 208, 225, 451, 532
Cincinnati Screw and Tap Co., 208
Cincinnati Shaper Co., 452
Clement and Sharp, 287
Clerm & Morse, 276
Cleveland, 437
Colt Armory, 253
Computing-Tabulation-Recording (IBM), 204
Cone Drive, 389
Corliss Co., 22, 251–52
Corning Glass Works, 498
Craven Brothers, 274
Crofts Engineers Ltd., 411
Crown Gear b.v. Holland, 530
Cunningham Industries, 520
C. W. Hunt, 383

D

Daimler Co. Ltd., 465
Dana, 362
Darling, Brown, and Sharpe, 225
David Brown and Sons, 409
De Laval Steam Turbine Co., 507
Derihon Steel Works, 363
Detroit Axle Co., 439
Detroit Emery Wheel Co., 231
Digital Electronics Automation, 521
D. O. James, 474
Dow Chemical Co., 498
Dr. Nott's Novelty Works, 218
Durand, 27, 447–48

E

Earle Gear Company, 341
Eastman Kodak, 327
Eberhardt, 26, 274, 421, 432
Edward G. Herbert Ltd., 376
E. G. Wrigley, 336
E. I. Du Pont deNemours Inc., 372
Elektrostal Heavy-Machinery Works, 445
EMAG Corp., 529
Enfield Co., 224

F

FAG Bearing Co., 405
Fagersta Iron Steel Works, 363
Falk Co., 385, 450
Farquarson A. B. Co., 348
Farwell, 27, 433
Fässler, 457, 467
Faulconer and Norton Co., 231, 236, 460–61
Fawcus Machine Co., 386
Fellows Gear Shaper, 268, 484
Ferranti, 487
Fischer A. G., 234
F. Lewis, 244
Flexible Laser System Co., 525
FNUC, 526
Foote, 253
Ford Motor Co., 415, 445, 508
Fox, 215
Frenco GmbH, 343
Friedrich Krupp A. G., 377

G

Gage, Warner, and Whitney Co., 274
G. A. Gray Co., 208
Ganschow Gear Co., 371, 388
Gardner Machine Co., 238, 462
Gaudard Lucien & J. D. Gibbs Co., 184
Gay & Silver Co., 215, 222
GCA Industrial Systems Group, 526
Gear Grinding Co., 461
General Electric Co., 319, 360, 372, 377–78, 400, 498, 532
General Motors Corporation, 335, 354–55, 360, 362, 389, 509, 525, 528
George C. Hagan Co., 525
Gisholt, 516
Gleason-Pfauter Machine Fabrik GmbH, 529
Gleason Works, 253, 255, 275, 359, 393, 395–96, 414, 422, 426, 439–40, 442, 478, 522, 525, 530
Goodhue Wind Engine Co., 20, 166
Gorham International Inc., 369

Gould and Eberhardt, 26, 274, 421
Grant, 267
Great Northern Paper Co., 372
Greenwood and Batley Ltd., 465

H

Hans Renold Ltd., 465
Hartness, 268
Hart-Parr Co., 348
Hawker Siddley Co., 411
Heald Machine Company, 465
Hendey-Norton Machine Co., 213–14, 412
Henry Wallwork Co., 472
Herbert & Fletcher Ltd., 376
H. Ernault Co., 438
Herschel W. H., 505
Hewes & Phillips Iron Works, 310
Hillman, 25, 336
Hinsley Gear Co., 388
Hofler, 484
Holmes Co., 255
Holtzappfel & Co., 241
Horsburgh and Scott, 341
Houghton International, 191
H. Siddley & Co, 411
Hudson Motor Car Co. of America, 376
Humber, 325
Humpage, Thomas, and Hardy, 432
Hupp Motor Car Co., 353

I

I. G. Farben Industrie, 498
Illinois Gear, 521
Illinois Tool Works, 396, 485–86
IMT-Fette Inc., 455
Intel, 517, 519, 521, 523
International Business Machines Corp., 204

J

Jackson-Church-Wilcox Co. (later Saginaw Div. Gen. Motors), 389
James Herron Laboratories, 345
J. E. Reinecker, 264, 272, 279, 446, 461
John Holroyd and Company, 283
Jones and Lamson, 268

J. W. Putnam, 236

K

Kapp GmbH, 457
Keller, 516, 532
Kemble Co., 218
Kerney & Trecker, 520
Keystone-Hindley Gear Co., 388
Klingelnberg & Sohne, 394, 443
Klockner, 360
Kymi Kymmene Metall, 361

L

Lamson Co., 268
Lanchester Automotive, 388
Landis Tool Company, 465
Lapointe Machine Tool, 444
Lawson and Cotton, 259
Leblond Co., 208
Leland, Faulconer, and Norton Co., 236
Lepel Corp., 360
Liebherr, 438, 529
Lincoln George S. Co, 224
Lincoln Motor Co., 235, 376
Lodge and Davis, 208
Lohmann and Stolterfoht GmbH, 507
London Emery Works Co. Ltd., 465
Ludwig Loewe Co. A. G., 374
Lumen Bearing Co., 368
Lumsden Machine Co. Ltd., 465

M

Maag, 28, 427, 466
MacConnell-Kennedy, 206
Mars Works, 240
Matlock Co., 488
Matthew Murray Works, 144, 149, 215, 220
Maudslay, Sons and Field, 151
Mazak, 528
M. Chambon, 432
Melville MacAlpine Co., 476
Mesta Machine, 268
Michigan Tool, 456
Midvale Steel Works, 193, 451
Modul Co., 526

Moore and Wright, 487
Morris Motors, 527
Motorenwerke, 507
Motorola, 343–44

N

National Broach and Machine, 456–57
National Tool Co., 455
National Twist Drill and Tool Co., 376
Newall Mfg. Co. Ltd., 465
Newark Gear Co., 341
Newmeyer Metal Works, 379
New Process Gear Corporation, 370, 407
New Process Rawhide Co., 370
Newton Machine Works, 414
New York Belting and Packing Co., 229
Nexen Group, 43
Nichols and Shepard Company, 509
Norton Company (Norton Emery Wheel Co.), 236
Norton Emery Wheel Company, 231, 461
Norton Grinding Co., 236, 460
NV Belting and Packaging Co., 229

O

Oerlikon Machine Works & Buehle Co., 394, 442–43, 462, 470
Oilgear Co., 444
Opel, 325
Osram Group, 377

P

Packard Motor Co., 512
Panhard, 186, 326
Parkinson Joseph Co., 21, 226
Parsons Corporation, 519
Peerless Co., 325
Pfauter, 264, 432, 448, 521, 529
Philadelphia Gear, 341
Pitcher & Brown Co., 234
Pittsburgh Gear, 341
Pope Co., 321
Power Plant Co., 477

Pratt and Whitney, 23, 201, 214, 219, 224, 253, 260–62, 278–79, 283, 444, 456, 505, 509
Providence Machine Tool Co., 243
Pullman Co., 251

R

Rambler, 325
R. D. Nuttall Co., 341, 383, 386, 439, 510
Red Star Windmills, 166
Reichraming Steel Works, 192
Reischauer, 463
Remington-Rand, 517
Renk, 162
Rennie Bros., 287
Riehlé Bros.-Testing M/C Co., 504
Riley, 325, 473
Robbins and Lawrence, 212–13, 217, 222
Roberts & Co., 287
Robey and Smith, 261
Rockwell Truck & Axle Division, 526
Rolls Royce, 205
Rover, 325
R. Sterne and Co. Ltd., 465
R. W. Meyer Co., 330
Ryder Company, 505–6

S

Samputensil S.p.A., 529
Sandia Labs, 533–34
Schuchardt and Schütte, 433
Shell Co., 506
Shin-Kobe Electric Machinery Co., 373
Siemens, 184, 188
Simonds Manufacturing, 341
Singer Mfg. Co., 484
Sir W. G. Armstrong Whitworth & Co. Ltd., 464
Sir W. H. Bailey and Company, 503
SKF, 405
Smith & Sandland, 366
Smith and Coventry, 261
Societe Anonyme Adolphe Saurer, 478
Spicer, 417
Spire Corp., 417
Standard Oil Co., 505

St. Joseph Iron Co., 244
Sunderland, 23, 27, 256, 427–28
Swasey Co., 263
Sykes, 430

T

Taylor-Hobson, 487
Taylor's Gear Grinding M/c Co., 461
Taylor-White Co., 373
Texas Instruments, 517
Thompson Mfg. Co., 163
Thurston Co., 197
Timken, 353
Timken-Detroit Axle Co., 439
Timken Engineering Surfaces Business, 381
Tinius Olsen, Inc., 197
Tool Steel Gear and Pinion, 379
Trecker Co., 520
Troughton and Simms Co., 202

U

Unimation, 525
United Calibration Corp., 366
Universal Radial Drill Co., 208
US Steel Corporation, 361

V

Vacuum Oil Company, 491
Van Baerle and Co., 231
Van Dorn Dutton Co., 341
Vickers, 366, 395

W

W & T. Avery, Ltd., 513
Walker Grinder Co., 465
Ward and Taylor Grinding Machine Co., 461
Warne and Co., 229
Warner Gear Co., 354
Westdeutsche Getriebewerke GmbH, 411
Western Company, 506
Westinghouse Electrical Co. and Manufacturing, 184
W. F. and John Barnes Company, 209
White Motor Co., 355
Whitworth and Co. Ltd., 267
Wilcox Corporation, 389
Wilfred Lewis Plant, 340
William Cramp and Sons Ship and Engine Building Co., 367
William Sellers and Co., 313
Williamson D. T. N. and Son, 533
Willys, 325, 335
Wilmarth and Morgan, 465
Winton, 325
Wright Bros., 326
Wyman & Gordon, 364

Y

Yale and Towne, 471

Z

ZIS Automobile Works, 359